Quality in Chemical Measurements

Springer-Verlag Berlin Heidelberg GmbH

Bernd Neidhart · Wolfhard Wegscheider (Eds.)

Quality in Chemical Measurements

Training Concepts and Teaching Materials

Springer

Editors:

Prof. Dr. Bernd Neidhart
GKSS Forschungszentrum Geesthacht GmbH
Institut für Physikalische und Chemische Analytik
Max-Planck-Straße
D - 21502 Geesthacht, Germany
e-mail: bernd.neidhart@gkss.de

Prof. Dr. Wolfhard Wegscheider
Montanuniversität Leoben
Institut für Allgemeine und Analytische Chemie
Franz-Josef-Straße 18
A - 8700 Leoben, Austria
e-mail: wegschei@unileoben.ac.at

ISBN 978-3-642-63016-3

Library of Congress Cataloging-in-Publication-Data
Quality in chemical measurements : training concepts and teaching materials /
Bernd Neidhart, Wolfhard Wegscheider, eds.
   p. cm
Includes bibliographical references.
ISBN 978-3-642-63016-3   ISBN 978-3-642-56604-2 (eBook)
DOI 10.1007/978-3-642-56604-2
Additional material to this book can be downloaded from http://extras.springer.com.

1. Chemistry, Analytic--Quality control--Study and teaching. I. Neidhart, Bernd, 1941 -
II. Wegscheider, Wolfhard.
QD75.4.Q34 Q238 2001
543'.071'1--dc21

This work is subject to copyright. All rights are reserved, whether the whole or part of the material is concerned, specifically the rights of translation, reprinting, reuse of illustrations, recitations, broadcasting, reproduction on microfilm or in any other way, and storage in data banks. Duplication of this publication or parts thereof is permitted only under the provisions of the German copyright Law of September 9, 1965, in its current version, and permission for use must always be obtained from Springer-Verlag. Violations are liable for prosecution under the German Copyright Law.

© Springer-Verlag Berlin Heidelberg 2001
Originally published by Springer-Verlag Berlin Heidelberg New York in 2001
Softcover reprint of the hardcover 1st edition 2001

The use of general descriptive names, registered names trademarks, etc. in this publication does not imply, even in the absence of a specific statement, that such names are exempt from the relevant protective laws and regulations and therefore free for general use.

Product liability: The publisher cannot guarantee the accuracy of any information about dosage and application contained in this book. In every individual case the user must check such information by consulting the relevant literature.

Typesetting: cameraready by editors
Cover design: design & production, Heidelberg
Printed on acid free paper   SPIN: 10664262   52/3020 - 5 4 3 2 1 0

> A sector which has always prided itself on its enduring commitment to high quality and its recognized competence has difficulty in accepting that henceforth it will have to demonstrate proof of quality and, moreover, invest in expensive measures to maintain it (Neidhart 1996).

Preface

Analytical data influence our daily life and nowadays criteria for assessing quality of chemical measurements must be classified as socio-politically relevant; thus Analytical Chemistry becomes part of general education.

The concepts of Analytical Quality Assurance (AQA) and Analytical Quality Management (AQM) developed in the wake of the harmonization of the European market and in connection with the globalization of the world's major trading zones, have now been formally established via the appropriate directives and norms (ISO 25, EN 45001 and recently ISO 17025). Although these developments have become widely accepted as market-regulating elements by both the chemical industry and independent laboratories for routine chemical analysis and are now practised extensively in the form of accreditation, this has taken place without any perceptible participation on the part of the universities. This state of affairs is somewhat similar to the situation prevailing at the beginning of the 80's with regard to the introduction of Good Laboratory Practice (GLP). The university sector which has always prided itself on its enduring commitment to high quality and its generally recognised competence appears to be having difficulty in accepting that henceforth it will have to demonstrate proof of quality and, moreover, invest in expensive measures to maintain it. In contrast to the situation with regard to GLP, this will be disadvantageous in the short to medium term to those institutions in the higher education sector that persist with their traditional structures and teaching curricula and thus fail to react to the developments in Analytical Chemistry which have given it the status of an independent scientific discipline with an increasing global-economic and socio-political importance.

In parallel to the development of the AQA concept, criteria for defining analytical quality have been conceived to permit comparability of analytical results. Use of these criteria enables comparability to be achieved via the traceability of analytical results to national or international standards along an unbroken chain of comparisons. Within the framework of AQA it is essential to be able to identify unequivocally the corresponding sample to which such high quality analytical measurements pertain (trackability?). Validation continues to be the central task in the development of any analytical method whose analytical capabilities in specific applications can be estimated with the aid of measurement uncertainty. Finally, proficiency testing serves to demonstrate comparability in terms of the scatter of the results, e.g. in round-robin tests.

In considering the role of AQA in the higher education sector it is necessary to differentiate between the various university activities which include services, research and development and teaching, as follows:
- Routine chemical analyses (including ad hoc analyses) performed for external clients and for the university's own measurement campaigns (e.g. investigations of the quality of waste-water and air) requiring full documentation.
- Routine chemical analyses carried out for internal clients as a service to research in other Chemistry Departments such as Inorganic, Organic, and Biochemistry.
- Chemical analyses performed as part of research and development work not only in Analytical Chemistry but also in other chemical disciplines such as Inorganic, Organic and Biochemistry.
- Chemical analyses carried out within the framework of research projects having pre-eminent goals which are analytically-based (e.g. studies of the temporal and spatial variations in metal-species concentrations in riverwater; determination of the gas composition in a waste incinerator as a function of the operating parameters).

These considerations also apply to the whole range of scientific disciplines in which chemical measurements are made, such as Biology, Geology, Medicine, Microbiology, Mineralogy, Ecology, Pharmacy, Toxicology etc.

Clearly, accreditation and GLP alone cannot guarantee that the results of chemical measurements are of high quality. This is because we can only assure (in the sense of AQA) that which exists or is produced and, of course, the development and validation of analytical methods are activities which require technical expertise. A direct consequence of this is that the universities will only be able to adapt to this new situation if they generate the prerequisites to permit the teaching of the modern concepts and strategies of Analytical Chemistry. This of course applies primarily to university teachers of Analytical Chemistry; however, university teachers from other disciplines wishing to take on this task will have to be self-taught.

In fact, it is very simple to integrate aspects of Analytical Chemistry into the basic practical work of an Inorganic Chemistry course with a minimum of additional resources and without any extra investment in time. Of course such innovations require that one is prepared to become involved with the topic! The introduction of new ideas is associated with an investment in time and effort such that in this area progress can only be made with difficulty.

At this point in time where the introduction of accreditation/recognition in the Higher Education sector is imminent, it is already clear that there are far too few academics having the necessary qualifications to carry out its implementation. For this and other reasons - in particular the growing economic and socio-political significance of Analytical Chemistry as referred to in the introduction - the teaching of this subject must be expanded.

This leads to the conclusion that:
- The quality of chemical measurements (or AQA) must become a sustaining element of modern research and teaching in the Chemistry departments of universities.
- Concepts to meet this demand will have to be developed. The prerequisite for this is an improvement in the teaching and training in Analytical Chemistry which can be achieved via
  - changes in the content, concepts and organization of teaching in the foundation, undergraduate and graduate courses in Chemistry,
  - the introduction of Analytical Chemistry as a compulsory subject,
  - intensive support for the new generation of academics in Analytical Chemistry
- Academic freedom in teaching and research also involves a responsibility
  - to adapt oneself to changed conditions,
  - to prepare students for new tasks,
  - to face the competition from other universities,
  - to give priority to fulfilling teaching duties, if necessary, at the expense of one's own scientific interests.

The 2nd EURACHEM Workshop on Current issues in teaching quality in chemical measurements held at GKSS (27-29 September 1998) enabled approximately 50 experts from 14 European countries to meet and exchange ideas on concepts for teaching quality with the aim to fill the gap between theory and reality.

The output of the workshop is published in this text book, comprising a collection of transparencies on CD-ROM. The small copies at the end of each contribution are for identification purposes only. It is hoped that the availability of this material will assist in reducing the "activation barrier" associated with the preparation of lectures and seminars on the topic.

Acknowledgements

The Editors appreciate the contributions and cooperation of the authors in revising and making up the original lectures. Thanks are due to Enno Bahlmann (GKSS Geesthacht) for the really hard work on the standardization of the transparencies and on the text processing and lay out. We also thank Marcus Krapp (Measurement Standards Laboratory of New Zealand) for the revision of the first draft and his extensive proof reading and correcting. Finally, our deepest thanks go to Heide Neidhart (GKSS Geesthacht) for her constant and untiring efforts in coordinating the submission of the authors' manuscripts and dealing with all the text compatibility problems inevitable in our modern information technology age.

Geesthacht and Leoben, June 2000

Prof. Dr. Bernd Neidhart

Institut für Physikalische und
Chemische Analytik GKSS-
Forschungszentrum Geesthacht
Max-Planck-Straße
D-21502 Geesthacht
(bernd.neidhart@gkss.de)

Prof. Dr. Wolfhard Wegscheider

Allgemeine und Analytische Chemie
Montanuniversität Leoben,
Franz-Josef-Straße 18,
A-8700 Leoben
(wegschei@unileoben.ac.at)

List of Contributors

Albus, Heinz-Erwin
Comdisco Deutschland GmbH
Oskar-Messter-Straße 24
D-85737 Ismaning
healbus@comdisco.com

Evans, E. Hywel
University of Plymouth
Dept. of Environmental Sciences
Drake Circus
Plymouth PL4 8AA, UK
hevans@plymouth.ac.uk

Green, John, D.
BP Chemicals Limited
Britannic House
Flinsbury Circus
London EC2M 7BA, UK
greenj2@bp.com

Houlgate, Peter
Laboratory of the Government Chemist
Queens Road
Teddington
Middlesex,TW11 OLY, UK
prh@lgc.co.uk

Koch, Michael
Institut für Siedlungswasserbau,
Wassergüte- und Abfallwirtschaft
Universität Stuttgart
Bandtäle 2
D-70569 Stuttgart
michael.koch@iswa.uni-stuttgart.de

Lee, Elliot
(via Peter Houlgate)
Laboratory of the Government Chemist
Queens Road
Teddington
Middlesex,TW11 OLY, UK
prh@lgc.co.uk

Pyell, Ute
Philipps-Universität Marburg
Fachbereich Chemie
Hans-Meerwein-Straße
D-35032 Marburg
pyell@ps1515-chemie.uni-marburg.de

Radvila, Peter
Neue Schweizer. Chemische Gesellschaft
Sektion Analytische Chemie
Rösslistraße 27
CH-9056 Gais
peter.radvilla@empa.ch

Ríos, Angel
Universidad de Córdoba
Departamento de Química Analítica
E-14004 Córdoba
qa1ricaa@uco.es

Rudd, David
Glaxo Wellcome
Analytical Science Department
Park Road
Ware
Hertfordshire SG12 ODP, UK
drr1605@glaxowellcome.co.uk

Steck, Werner
BASF AG,
ZAX/A-M320
D-67056 Ludwigshafen
werner.steck@basf-ag.de

Townshend, Alan
University of Hull
Chemistry Department
Cottingham Road
Hull, HU6 7RX, UK
a.townshend@chem.hull.ac.uk

Valcárcel, Miguel
Universidad de Córdoba
Departamento de Química Analítica
E-14004 Córdoba
qa1meobj@uco.es

Wampfler, Bruno
Eidgenössische Materialprüfungs- und
Forschungsanstalt (EMPA)
Lerchenfeldstraße 5
CH-9014 St Gallen
bruno.wampfler@empa.ch

Wegscheider Wolfhard
Allgemeine und Analytische Chemie
Montanuniversität Leoben,
Franz-Josef-Straße 18,
A-8700 Leoben
wegschei@unileoben.ac.at

Contents

IMPORTANT INFORMATION FOR READERS AND USERS OF THE CD-ROM — XIII

IMPORTANCE OF ANALYTICAL QUALITY MANAGEMENT AND QUALITY ASSURANCE IN INDUSTRY, ACADEMIA AND RESEARCH PROJECTS

Why do we need Good Results? — 3

The Importance of 'Good' Measurement on Industrial Manufacturing Efficiency and Profit — 11

Concepts of Quality Management and Quality Assurance in Analytical Research Projects and Non Routine Analysis — 23

WORKED EXAMPLES OF TEACHING ANALYTICAL QUALITY CONCEPTS

Quality Systems for Non-routine and R&D Analytical Work - Accreditation of Non-routine Laboratories — 33

Evaluation of Uncertainty in Analytical Measurement — 43

Traceability / Trackability — 65

Validation: an Example — 79

Metrology in Chemistry — 89

EXPERIMENTS AS TOOLS TO DEMONSTRATE PRINCIPLES OF QUALITY ASSURANCE

Basic Course Experiments to Demonstrate Validation — 111

Basic Course Experiments to Demonstrate Intercomparisons — 121

Assessment of Test Kits in Terms of Time, Cost and Quality — 129

Estimation of Random Deviations in Analytical Methods using Analysis of Variance — 143

COURSE STRUCTURES, CONTENTS AND EXPERIENCES

PT Scheme for Pre-university Students — 151

Teaching of the Concept of Valid Analytical Measurement: Integration of Quality Assurance (QA) Issues or Separate QA Courses for Higher Education — 163

Special Requirements for Interlaboratory Proficiency Tests — 171

Important Information for Readers

Viewing and printing the transparencies:
The transparencies on the enclosed CD-ROM are edited in Microsoft® PowerPoint® 97. If you do not have Microsoft® PowerPoint® 97 on your PC you can look at and print the transparencies with Microsoft® PowerPoint Viewer 97, which is available as shareware on the Microsoft homepages.

Important notice:
You may access the documents containing the transparencies either via opening the CONTENTS.DOC file on the CD and clicking on the respective hyperlinks or by viewing the directory \TRANSPARENCIES on the CD (e.g. by using Microsoft® Explorer) and clicking on the respective file names (*.ppt).
When printing the overheads, please remember to set your printer to the right settings, regarding e.g. the medium of output (paper, overheads, ...), colour, size. If necessary, please consult your PowerPoint® and/or printer handbook.

System Requirements:
- IBM PC or compatible with a 486 or higher processor
- Microsoft Windows® 95 operating system or Microsoft WindowsNT® Workstation operating system 3.51 with Service Pack 5.x or later
- RAM: 4 MB (Windows 95) (8 MB recommended), 12 MB (Windows NT)
- 13 MB of hard disk space (15 MB free for installation only)
- VGA or higher-resolution video adapter
- Microsoft® PowerPoint Viewer 97 (shareware) or PowerPoint® 97 (recommended)

Copyright and License:
1. The data stored on the CD-ROM are protected by copyright. Any rights in them lie exclusively with Springer-Verlag.
2. The user may use the transparencies, print-outs thereof and multiple copies of the print-outs in classrooms and lecture halls. A l l copies must show the copyright notice of Springer-Verlag.
2. The user is entitled to use the data in the way described in section 2. Any other ways or possibilities of using the data are inadmissible, in particular any translation, reproduction, decompilation, transformation in a machine-readable language and public communication; this applies to all data as a whole and to any of their parts.

Liabilities of Springer-Verlag
1. Springer-Verlag will only be liable for damages, whatever the legal ground, in case of intent or gross negligence and with respect to warranted characteristics. A warranty of specific characteristics is given only in individual cases to a specific user and requires an explicit written representation. Liability under the product liability act is not affected hereby. Springer-Verlag may always claim a contributory fault on the part of the user.

2. The originator or manufacturer named on the product will only be liable to the user, whatever the legal ground, in case of intent or gross negligence.

ADDITIONAL CONDITIONS FOR USERS OUTSIDE THE EUROPEAN COMMUNITY: SPRINGER-VERLAG WILL NOT BE LIABLE FOR ANY DAMAGES, INCLUDING ANY LOST PROFITS, LOST SAVINGS, OR OTHER INCIDENTAL OR CONSEQUENTIAL DAMAGES ARISING FROM THE USE OF, OR INABILITY TO USE, THE ACCOMPANYING CD-ROM AND ITS CONTENTS, EVEN IF SPRINGER-VERLAG HAS BEEN ADVISED OF THE POSSIBILITY OF SUCH DAMAGES.

Importance of Analytical Quality Management and Quality Assurance in Industry, Academia and Research Projects

Why do we need Good Results?

J. D. Green

Abstract
Industrial Chemistry is involved with the production of chemicals to make a profit. Any process must be commercially viable to survive. However, whilst this is a necessary requirement, it is not sufficient. Chemical production must produce materials of the required specification, they must be produced safely and lawfully and the environment should suffer no harmful effects. Forward-looking companies need to be involved in research and to be aware of new developments so that improved production methods can be introduced. This leads to the requirement for commercially sensitive information to be protected.

All the above activities are to some degree dependant upon analytical chemistry and quality in industrial analytical chemistry is essential if the various aspects of production operations are to be achieved successfully.

It is necessary to recognise that there are different aspects of analytical quality. Quality requirements depend upon the end use of the analysis whether that be ensuring the efficiency of a production operation or protecting the environment from any fugitive emissions.

A teaching lecture relating to the above topic should:
- Establish by the use of examples why the different aspects of responsible industrial chemical production depend upon analysis.
- Define different aspects of analytical quality (eg - accuracy, repeatability, timely).
- Describe why analytical quality is related to the end-use to which the results are to be put.
- Establish the advantages of good quality results.

BP Chemicals produces a range of bulk chemicals and polymers at several sites throughout the world. The Hull site specialises in ethanoic acid and related organic acids, a range of esters and phthalic anhydride and related esters. Raw materials are brought on to site from a variety of suppliers and products are supplied to a wide range of customers in many different industry sectors. This description would have similarities to other chemical manufacturing facilities throughout the chemical industry.

Slide 1
Underlying the production of particular chemicals, such as ethanoic acid, there are guiding principles that again would be similar across the chemical industry. Chemicals must be produced profitably, it would be impossible for the industry to survive if the chemicals it produced could not be sold at an economical price. This implies that the materials are produced to the correct specification, a specification that makes them attractive to customers. The products should be produced without

harm to the environment - these days companies must operate with the consent of the local population and those that don't have a limited lifetime. Operations must be safe so that no one is injured in the manufacturing operations. In other words the operations must be conducted lawfully from a health safety and environmental point of view. Furthermore chemical companies must protect their intellectual property and observe the rights of others in this area. All of these requirements are assisted by the use of good analytical procedures that provide good results.
(Students could be asked for ideas as to the objectives of chemical companies.)

Slide 2

However, what are 'good' results? A good result has to be viewed in the context of the requirement for that result. This is usually summed up by the phrase 'fit for purpose' meaning that the result provides the data and thereby the information that is required. In many cases this implies that decisions can be made on the basis of the results. Results sometimes need to be quantitative and sometimes need only to be qualitative. Some results will be accurate, others precise. However, results invariably depend upon the analytical methods that produce them, methods can need to be valid, repeatable, reproducible and consistent. And then having produced the results they need to be provided in the appropriate form, on time, obtained with the minimum effort and cost. It is often important to store results in an audited and retrievable form for if this is not the case they can at a later stage be found to be worthless. A good result is fit for purpose is traceable and is generated from a validated method.
(Students are asked about 'Good' results - What makes a good result?)

Slide 3

Although industry is principally concerned with successful manufacture research underpins future developments and 'supplying the customer' must be an integral part of the operation. Health, safety and the environment are in most organisations rightly regarded as essential elements of a successful operation. Analysis plays its part in all these activities.

Slide 4

If one considers industrial research in the chemical industry it soon becomes obvious that measurement is vital and analytical chemistry is an essential component of the measurement tools needed. Furthermore 'good' results, generated using well authenticated methods carried out in an efficient way make the difference between good and poor research. As an example one may consider the conversion of butane to ethene - a dehydrogenation reaction - important as a model cracking process to produce an important feedstock for chemical processes. The dehydrogenation can result in several products. The success of the process depends upon an efficient conversion to the required product and the suppression of the other by-products. Accurate analysis is required in order to be able to assess the true conversion of butane to ethene. If the value obtained was inaccurate then any conclusions drawn as to the viability of scaling up the process would be invalid. Sensitivity is needed to ensure that any minor by-products are detected in the small scale research equipment before the process is considered for transfer to a production scale. Flexibility of analysis is always important in research projects because it is at this stage that unexpected requirements become apparent.

(Students can be presented with a simple reaction and asked to consider the consequences of having poor results, for example the inability to detect a component in a research project that is inert and forms part of a recycle in the eventual production plant.)

Slide 5
Once a process is an established manufacturing process the analytical requirements change. The production process should be able to be carried out repeatably time after time and the analytical monitoring of the process (sometimes referred to as process analysis and control) needs to be able to detect changes to the operational conditions so that action can be taken. Accuracy of the measurements is less important because fluctuations can be detected if the analytical procedure has a high degree of precision. It is often important to detect process changes quickly before off-spec material is produced, the frequency with which the analysis can be performed is thus very important. The analysis should also be economical taking into account the cost of the equipment and the cost of doing the analysis. And lastly when the results are obtained they must be displayed and fed back to the manufacturing operation so that they can be used effectively.

So, good results for research and development can be different from good results for a production process.

Slide 6
Customers of the chemical industry demand quality products which conform to the specifications of the manufacturer. For this to be so accurate analytical results are required with which the manufacturer and the customer can be confident. A good analytical result for this purpose must be obtained by a method which is recognised by all and has a demonstrated validity. The result obtained in one laboratory should be comparable with the result obtained elsewhere.

Slide 7
Analytical techniques have a bearing upon the quality or purity of products. Materials required at high purities cost more to produce. This can be readily understood by considering the purification of a solid material by fractional recrystallisation. The higher the purity required the more steps needed in the process and therefore the more time used. And secondly in purifying a liquid by distillation higher purity often requires a lower take-off from the top of the distillation column so more time is again required and more heat is used in keeping the vapour liquid equilibrium of the column. However is can be shown that a better analytical method - in this case one with higher precision - can actually reduce the cost of obtaining high purity material. This is illustrated in the next slide.

Slide 8
It is important here to summarise briefly and simply the underlying principles of analytical confidence limits. All measurements have errors associated with them and these can be expressed in terms of a confidence interval, for example $\pm 1\%$. Analytical values obtained with that technique are then expected to lie within that interval so anyone determining a value of 17% would be confident that the actual value was within the range 16 to 18%. This confidence interval must be

considered when a technique is used to ensure conformance to a given value, for example a specified purity value.

Slide 9
The curve represents the increasing cost of producing very pure material. As the purity needs become greater then the cost of purity increases. Let us assume that a purity of 98% is essential and our analytical technique gives a value of $\pm 0.5\%$ then, in order to be confident of producing material of the correct specification, an analytical result of 98.5% must be obtained. Such a purity will be produced at a specified cost. If the analytical method can be improved to provide a better result, for example $\pm 0.1\%$, then a result of only 98.1% has to be achieved in order to be confident of producing material of the 98% purity required. Again the cost of achieving the 98.1% purity result is specified and there is clearly from the figure a cost saving. Better analysis therefore translates into cost savings in this case.
(Students could be provided with a table of cost versus purity figures together with two different analytical techniques which had different error bands. From theses figures they would be asked to calculate the cost savings achieved by using the better analytical result.)

Slide 10
Good analytical results allow chemical plant to be used more efficiently. This contributes towards the effective economic use of the assets. This is clearly desirable because a chemical plant is expensive. The next two slides illustrate how both batch processes and continuous processes benefit from the use of good analytical results.

Slide 11
In a typical batch reaction one or more of the reactants can be monitored and completion of the process is indicated when the concentration of one component drops to a specified level. The primary products are then removed from the batch reactor and either sold as produced or purified by further treatment. The equipment can only be reused when it has been emptied and re-prepared for the next batch. It is therefore desirable to attain maximum usage of the equipment and to do this requires the batch time to be reduced as much as possible. The slide shows the reactant concentration reducing towards and then dropping below the value indicating the required conversion. With the first analytical technique to be sure of the conversion needed the analytical value obtained must drop considerably below the conversion value required. With the improved technique the reactant concentration does not have to drop so much before the process is regarded as complete. The difference in the times involved represents a considerable saving on the batch time and therefore on the asset utilisation.
(This is an example where students could be given a table of figures relating to the progress of a batch reaction together with the errors of two different analytical techniques. The exercise would involve calculating the time saved in using the better technique.)

Slide 12
For a continuous process the slide shows the function relating throughput of the chemical process plant plotted against the purity of the product. As the product purity gets higher the throughput becomes less. For any chemical manufacturing process there will be a specified purity required, in this example it is given as 98%. Two analytical techniques are considered the first having a greater error than

the second. In each case to be sure of attaining the specified product purity a certain analytical result must be delivered. The slide shows that the better analytical technique, delivering better analytical results, gives a increased throughput and therefore is helping to utilise the assets better. For a large chemical plant, typically producing hundreds of thousands of tonnes per year, increases of 1% can have a significant effect upon the economics of the process.

Slide 13

I have largely dwelt upon the economic benefits of using good results however there are indirect benefits which are as important to the long term prosperity of the chemical industry. These focus upon the use of analytical results for maintaining and improving the company's performance in the areas of health safety and the environment. Emissions and waste disposal have impacts upon the environment. Generally a reduction in emissions and disposals is desirable. Reliable analytical results obtained rapidly can avert a problem and can reduce the amount of material lost to the environment from chemical processes. However, good analytical results here can almost always be better and this provides a number of challenges. Techniques to determine lower concentrations can warn of small leaks that if unchecked could lead to major spillages. Techniques that can provide results from on-line measurements avoid sampling exposure of personnel. Techniques that are sensitive can be used to monitor effluent streams frequently and therefore detect fluctuations to normal operations. And techniques that can cope with contaminants without interfering with the result required can provide better results and improved information upon which to base actions.

Slide 14

Good analytical results detect corrosion before it becomes serious. Corrosion is one of the main factors leading to the necessity for chemical plant maintenance involving regular shutdowns, some unscheduled shutdowns and the replacement of components. The better the results obtained the easier it is to minimise the effects of corrosion thus increasing production efficiency and reducing the need for shutdowns.

Slide 15

So good analytical results are vital to industry but they are not only a cost upon industry they are also a means of increasing productivity, using assets more effectively and improving those performance measures which have a significant bearing upon a company's reputation. In summary there are a number of principles that those seeking to adopt good analytical practice and therefore good results should follow - they form the basis of the Valid Analytical Measurement programme - a programme spearheaded by the Laboratory of the Government Chemist in the UK.

      Valid Measurements should ...
          satisfy an agreed objective.
          use the correct methods and equipment.
          be carried out with appropriately trained staff.
          involve regularly assessed laboratories.
          be consistent wherever they are made.
          be made alongside defined QC and QA procedures.

J. Green Why do we need Good results? 1

Chemical Industry

- Purpose: To produce chemicals.....
  - to the correct specification
  - profitably
  - without harm to the environment
  - safely
  - lawfully
  - whilst protecting intellectual property

Analytical Chemistry has its part to play.....

J. Green Why do we need Good results? 2

Good Results

- Fit for purpose
- Accurate
- Precise
- Qualitative / Quantitative
- Consistent
- Allow decisions to be made
- Provided in an appropriate form
- Provided on time
- Obtained at minimum cost & effort
- Retrievable /Traceable
- Validated
- Repeatable
- Reproducible

J. Green Why do we need Good results? 3

Industrial Functions

- Manufacture
- Research & Development
- Supplying the customer
- Health Safety and Environment

J. Green Why do we need Good results? 4

Industrial Research

(Example)

Example - conversion of butane into ethene

- Possible dehydrogenation reaction

  butane ⟶ ethene + propene + butene
- Important parameters:
  - Conversion of butane into products
  - Selectivity of conversion to ethene
- Analytical requirements
  - accuracy ⇒ to ensure scale-up is valid
  - sensitivity ⇒ to detect low concentrations
  - flexibility ⇒ to identify unexpected products

J. Green Why do we need Good results? 5

Chemical Manufacture

Analytical Requirements

- precision of measurement
  - monitors progress of reaction
  - detects process fluctuations
- accuracy less important
- frequency of analysis
- minimum cost & effort
- display of results/feedback protocols

J. Green Why do we need Good results? 6

Supplying the Customer

- **Accuracy**

  ↳ensures product is 'in-spec'
- **Demonstrate validity**

  ↳a recognised method
- **Reproducibility**

  ↳comparable result whoever carries out the analysis
- **Consistent**

  ↳same sample same result

Slides 1-6

Analytical Techniques & Product Purity

- High purity is expensive
  - More steps required in fractional crystallisation
  - Less take-off in distillation columns
- Better analytical techniques reduce the cost of obtaining high purity
  - Greater precision allows required purity to be targeted more closely

Analytical Confidence Limits

- All measurements have errors associated with them
- Quoted values should therefore be expressed with an associated confidence interval [X ± x units]
- To be assured that a sample conforms to a quoted specification limit an analytical value must be obtained for the sample that takes the confidence interval into account.

Example:
- Technique gives value ± 1%
- Sample must be >97% pure
- Therefore technique must deliver a value of 98% to be confident of sample purity being >97% pure

Analytical Techniques & Cost of Quality

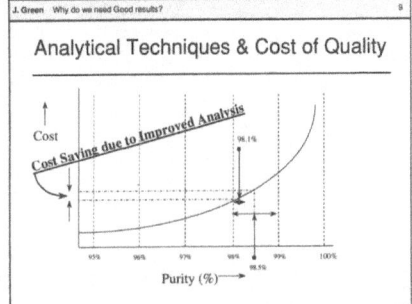

Analytical Techniques and Asset Utilisation

- Chemical plant is expensive
- Maximum utilisation is desirable
- Analytical techniques help to maximise the utilisation
  - Reduce turnround times on batch operations
  - Increase throughput on continuous plant

Analytical Techniques and Asset Utilisation
Batch Operations

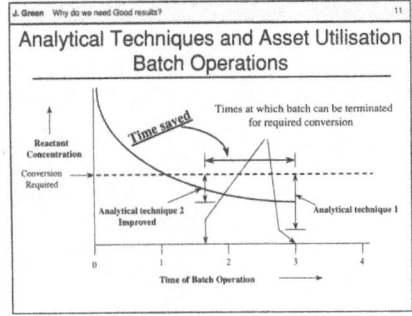

Analytical Techniques & Asset Utilisation
Continuous Processes

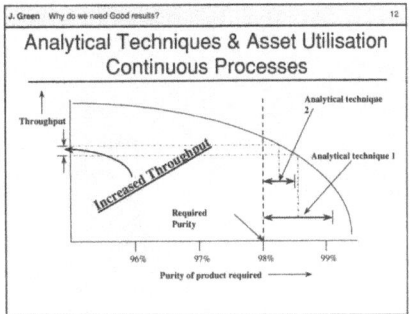

Slides 7-12

Analytical Chemistry & Environmental Responsibilities

- Emissions to Air
- Emissions to Water
- Waste disposal - off-site
- Technical challenges
  - low concentrations in air/water/solids
  - sample collection times
  - sensitive techniques required for temporal resolution
  - samples containing contaminants

Corrosion Measurements

- Chemical Plant Maintenance
  - Regular shutdowns / unplanned shutdowns
  - Components replaced
- Corrosion /Production /Shutdown Frequency
  - Relationship between corrosion & process variables .....
    - higher productivity whilst maintaining plant integrity
    - lower frequency of shutdowns

Valid Analytical Measurement Principles

- Analytical Measurements should be made to satisfy an agreed objective
- Analytical Methods should be made using methods and equipment which have been tested to ensure they are fit for purpose
- Staff making analytical measurements should be both qualified and competent to undertake the task
- There should be regular independent assessment of the technical performance of a laboratory
- Analytical measurements made in one location should be consistent with those made elsewhere
- Organisations making analytical measurements should have well defined quality control and quality assurance procedures.

Slides 13-15

The Importance of 'Good' Measurements on Industrial Manufacturing Efficiency and Profit

D. Rudd

Abstract
Assessment and assurance of quality within the field of pharmaceutical manufacture is established by the development of appropriate specifications for raw materials and products. The importance of suitable specifications will first be established and examples will be given of the development of a typical specification for a pharmaceutical drug substance. This will include discussion of the types of parameters which need to be controlled and an indication of the numerical values or ranges expected for most modern materials. The analytical methodology which needs to be applied to demonstrate compliance (or otherwise) with the specification will be discussed, along with the implications of method performance necessary in order to distinguish between variations in product quality and simple analytical variability.

A workshop session will also be included which will allow students to simulate the typical discussions which might take place in any large multi-national pharmaceutical organisation when establishing specification ranges or values for a modern drug substance based on a review of batch data generated during the development and early manufacture of the drug.

As a result, students will understand the importance of the control which appropriate specifications bring, the implications of the level of analytical method performance required and the combination of good science and pragmatism used by industry in developing such specifications and methodology for assurance of the quality of modern pharmaceutical drug substances and products.

Slide 1
Historically, there are a number of examples of man's activities which have proved either unsuccessful or, in extreme cases, disastrous and which are broadly attributable to a lack of control of key quality parameters.

The failure of the Piper Alpha oil rig in the North Sea in 1988, with an associated heavy loss of human life, and, no less dramatically, the several spectacular, yet abortive attempts by both US and Soviet astronauts to extend man's knowledge of space travel represent some of the more memorable examples of such lack of control and the consequences of this position from the safety perspective.

Within the field of medicine, the horrific thalidomide story of the 1960's probably represents the most famous example of the impact of inadequate control or specification of associated impurities in pharmacologically active materials.

Thus, the importance of appropriate control or specification of materials, whether they be of the structural or medicinal variety, becomes clear, not just from the point of view of safety or toxicity, but also in terms of assuring that the overall quality, however this may be defined, of the finished product is established at an appropriate and guaranteed level.

Slide 2

Within pharmaceutical development and manufacture, a wide variety of materials need to be subjected to an appropriate level of quality control. Most modern pharmaceutical preparations contain a number of raw materials or ingredients (in addition to the active drug substance) and each of these is subject to strict control or specification - largely governed by the expectations of the user or patient who, quite rightly, is protected by the Regulatory Agencies responsible for the marketing approval of the pharmaceutical product in the various commercial territories of the World.

Such control extends beyond just the simple constituents of the pharmaceutical preparation, however, and encompasses all components of the finished product (hence, includes packaging materials, activating devices etc) as well as the finished product itself.

Naturally, the nature of the controls which need to be applied varies according to the type of ingredient, the influence of the packaging materials and the form of the finished product but, nevertheless, appropriate controls are still required for each of these types of materials used during the development and manufacture of medicinal and other products.

Slide 3

A large number of different types of raw materials are used during pharmaceutical manufacture. These can generally be classified as 'excipients' -the pharmacologically inactive ingredients which are added to the active drug substance during manufacture and which may impart particular properties to the finished product (such as disintegrants in tablets, solubilising agents, flavours and colours etc) - or 'solvents and reagents' which might be used to facilitate the manufacturing process and which may, therefore, not be present as substantial components in the final composition of the finished product.

Appropriate control of all of these raw materials is necessary, however, in order to ensure that each is effective in terms of the desired action or properties it is intended to impart. Morever, the quality control of such raw materials is also likely to incorporate some limitation of the level of impurities in order to ensure that there is no toxicological risk to the user or patient as a result of such potential contamination.

Slide 4

Other components of the finished product include the packaging materials which are inevitably associated with the manufacture of pharmaceutical preparations. Such packaging materials serve a number of purposes, not least of which is the provision of protection of the active drug substance to the influences of heat, light or components of the atmosphere, for example. Thus, tablet products are often foiled in aluminium strips as these are extremely effective at preventing the ingress of atmospheric moisture, as are high density polyethylene bottles which may be used as bulk containers for many solid dosage forms.

Traditionally, of course, many pharmaceutical preparations are presented in glass containers (ampoules for injectable products, bottles for liquid presentations) and, once again, these afford excellent protection (as well as being a convenient form for dispensing) against a number of potentially adverse influences. Glass containers can even provide a high level of protection against the influence of light, provided a suitably coloured form of glass is used.

It is true, however, that the use of such packaging materials is not without its disadvantages as residual impurities present in the packaging components can migrate into the pharmaceutical preparation itself and can potentially become a source of adverse pharmacological reaction. Residual monomeric impurities from polyethylene could be one example, while even glass itself can allow the chemical degradation of the active drug substance as a result of local influences on the pH of the contents of the container.

Slide 5

The quality control of the finished products themselves clearly needs to be extensive as these are the materials to which the user or patient will be exposed most directly. Much of the control is directed at the use and intended form of administration of the finished product - thus, the control of tablet products will include a component which ensures that the tablet will disintegrate within an appropriate time interval, while injectable products will need to have a defined pH range to ensure patient tolerance etc - but, equally, emphasis will also be placed on ensuring freedom from (or, at least, limitation of) potentially toxic impurities. In short, finished product specifications will incorporate appropriate control parameters which will ensure both the efficacy and safety of the pharmaceutical preparation to a suitable level.

Slide 6

In order to illustrate in more detail the way in which pharmaceutical efficacy and safety can be assured through the use of appropriate specifications, emphasis will be placed on the control of the active drug substance. The principles applied, however, are identical for finished products, for raw materials and for packaging components.

To clarify what is meant by 'the active drug substance' - as opposed to the pharmaceutical preparation or finished product - examples are provided of some of well-known active drug substances and their associated finished products.

Thus, ranitidine hydrochloride is the active drug substance present in Zantac tablets, while salbutamol sulphate constitutes the active ingredient of Ventolin inhalers.

Slide 7

Although the term 'specification' is generally widely understood, within the sphere of pharmaceutical development and manufacture, a number of different types of specification are encountered. Thus, for clarity - and to emphasise the different facets of pharmaceutical manufacture - these various types of specification are described

In-process controls:

These are controls applied to the manufacturing process where measurements can be made in 'real time' and which are designed to ensure that the manufacturing process is under control. Examples of such in-process controls might include on-line weighing of individual tablets, monitoring of pH during bulk manufacture of an injection solution or temperature measurement during a fermentation process.

Release or manufacturing specification:
This is the quality specification which will be applied at the time of manufacture of the drug substance or finished product. On manufacture, the drug substance or finished product will be tested by the Quality Control laboratory and approved (or otherwise) against this specification.

End-of-life specification:
This is the quality specification which will apply throughout the life-to-expiry (often several years) of the drug substance or finished product. Closely related to the release or manufacturing specification, the end-of-life specification recognises that some deterioration in product quality is inevitable over the life-to-expiry of the product and, therefore, incorporates parameter ranges which reflect this change compared to the release or manufacturing specification. The end-of-life specification assures product quality, efficacy and saftey throughout the life of the product under the defined storage conditions.

Pharmacopoeial specification:
This specification encompasses all forms of the defined drug substance or finished product, rather than that applied by the originator or single manufacturer of the particular material. Thus, while Glaxo Wellcome, for example, might provide an end-of-life specification for Zantac tablets (the Glaxo Wellcome product), the pharmacopoeial specification would encompass any tablet product containing ranitidine hydrochloride as the active drug substance (of which 'Zantac tablets' are but one manufacturer's example).

Slide 8

A list of typical parameters which might establish quality for an active drug substance is provided.

Confirmation of *identity* is necessary in the first instance and the quantitative aspect which guarantees drug potency *(assay)* is included.

Levels of impurities (particularly the *drug-related impurities* which are likely to be responsible for any adverse toxicological effects) must be controlled. *Solvents* (traces of which may remain from the synthetic manufacturing process) and *heavy metals* (perhaps from residual catalysts) must also be limited by appropriate controls.

Finally, *particle size* is included as an example of the importance of control of physical parameters as well as chemical. Such a parameter could be extremely important, for example, for a drug which is likely to be administered in inhaled form, where penetration into the site of therapeutic activity is highly dependent on the aerodynamic diameter of the particle.

Slide 9

It is important when defining appropriate quality control specifications to distinguish between the inherent properties of a material (which are likely to be a feature of the molecule rather than a measure of batch quality) and those parameters which can be used to characterise successive batches of that material.

Thus, properties such as *solubility, chemical stability, density etc* are likely to be fairly consistent for a given material representing, as they do, features of that molecule. As measures of quality from one batch to another, these properties are unlikely to fluctuate to any great extent, making them relatively poor indicators.

Slide 10

So, having defined the likely indicators of batch quality (assay, drug-related impurity levels, solvent levels etc), it now becomes necessary to establish appropriate ranges or limits for these parameters.

Typically, drug substance specification ranges or limits reflect a mixture of scientifically-based judgement in terms of the suitability of the established specification and pragmatism with regard to the level of quality which is likely to be achievable during manufacture. Clearly it would be nonsensical to establish specification ranges or limits which cannot be met due to manufacturing limitations. Conversely, unnecessarily liberal control to reflect poor manufacturing capability is also inappropriate.

Thus, typical ranges and limits are presented which reflect appropriate control for most modern drug substances.

As assay range of 98 to 102% by weight is generally readily achievable with present day synthetic capability and is supported by limiting the total level of drug-related impurities (both synthetic and degradative) to no more than 2% of the nominal drug content.

In this example, solvents and heavy metals are also limited to typical levels (1% and 20ppm respectively).

Slide 11

Again, a list of typical analytical techniques used to measure and control the defined parameters is presented.

Infra-red spectroscopy and liquid chromatography (using a suitable characterised reference standard) are techniques which are widely employed within the pharmaceutical industry to confirm identity of materials.

Liquid chromatography, titrimetry or UV-visible spectroscopy are all useful techniques for establishing drug potency or assay, while the high level of selectivity and powerful quantitative aspects of liquid chromatography lend themselves ideally to measurement of levels of drug-related impurities.

Gas chromatography may be best-placed to allow monitoring of residual solvent levels, while trace inorganics species (heavy metals, for example) can be quantified and/or detected using atomic absorption spectroscopy, ion-exchange chromatography or simple semi-quantitative or qualitative colour reactions.

Finally, the physical attributes, such as particle size, may be successfully monitored and controlled using light scattering techniques or sieve analysis.

Slides 12, 13 and 14

Although the above techniques find wide usage in most modern pharmaceutical laboratories, a number of issues arise in terms of the required analytical method performance based on the intended application of the method and the likely specification range encountered.

As an example, consider the application of a high pressure liquid chromatographic assay method for the control of a drug substance.

Typically, such a method is likely to give an overall level of precision of the order of 1% RSD (for n=6, for example) when applied repeatedly to a drug substance sample.

Now, suppose that such a drug substance has a defined assay specification of 98 to 102% by weight. Is the performance of the analytical method sufficient to allow correct assessment of the quality of the chosen batch of drug substance?

The key question arises if, for example, an assay result of 98.1% is obtained. Does this indicate compliance of the batch with the defined specification or are more determinations required (recognising that any analytical measurement has an associated confidence interval which is related to the overall level of precision attributable to the method)?

In this example, it would be advisable to make additional measurements to confirm the suitability of the batch, although this practice would be unlikely to be well-received in most commercial quality control environments, where delays in batch release can have a serious impact on efficiency and productivity.

Consider, too, the analogous situation of a batch assaying at 97.9%. Does this constitute a batch failure or does it simply reflect the lack of precision of the chosen assay technique?

A natural conclusion to this discussion is that, for relatively narrow specification ranges such as drug assay, a high precision analytical technique is necessary.

Slides 15 and 16

Considerations of analytical performance also apply to control of levels of impurities, where the key aspect of the method centres around the sensitivity of detection and quantitation.

Regulatory agencies now insist that toxicological evaluation be carried out on any drug-related impurities which are present at levels at or above 0.1% of the nominal drug content. This has far-reaching consequences in terms of the levels of sentitivity required by most impurity methods in that detection and quantitation limits must clearly be considerably below this specification limit.

Slides 17 to 24

This sequence of slides provides an opportunity to simulate the review of typical batch data which might take place when attempting to establish an appropriate specification for a drug substance.

Data are provided for six batches of a drug substance and include assay, sodium, water, methanol and drug-related impurity measurements.

Key points which should be brought out during the discussions are as follows:

1. Taken in isolation, the assay data are unremarkable (but see Point 5 later)

2. Although some variation in sodium data is evident, no control on the upper limit may be necessary unless the sodium is present in hazardous or toxic form.

3. A water content limit of 'not more than 1%' might seem appropriate based on the batch data, but the only likely impact of water being present in the drug substance is one of hydrolytic instability. On that basis, a review of the drug

stability data (incorporating stability studies on batches containing different levels of water) might indicate that a revised (lower) limit is necessary. Conversely, if drug stability is satisfactory even with, for example, 3% water present, then an upper limit of 2.5% (for example) may be appropriate.

4. Consideration of the methanol data indicates that Batch C is significantly different to the other five batches. This is likely to be attributable to some atypical treatment of Batch C or to erroneous analytical data. The proposed specification limit should not attempt to encompass such atypical batches - rather the reasons for the unusual data should be investigated and understood.

5. From the review of the impurity data, it is evident that batch quality is better for the latest three batches (D, E and F) when compared to the first three (A, B and C). Thus, specifications for the principal impurity, the secondary impurity and the total level of impurities should probably be based around these later batches as these are more likely to reflect the quality achievable during routine manufacture.

Specification limits of 'not more that 0.5% for the principal impurity', 'not more than 0.3% for the secondary' and 'not more than 1.0% for the total' seem justifiable. There are some issues around mass balance, however, when the assay data and total impurity data for each batch are reviewed in conjunction with each other.

Even allowing for method variability, an assay of 101.0% together with total impurities of 1.5% for Batch C confirm its atypical nature and suggest that, somewhere, the analytical data is likely to be erroneous. On that basis, it should not be included in any deliberations designed to establish specification ranges or limits. In addition, a full investigation into the cause of the poor analytical results should be instigated.

Slides 25/26/27
Self explanatory summary

Importance of 'Good' Measurements on Industrial Manufacturing Efficiency and Profit

Importance of quality control

- Piper Alpha oil rig disaster in 1988
- US and Soviet space missions
- Thalidomide

Quality Control of Pharmaceuticals

- Raw materials
  (including the active drug substance)
- Packaging materials
- Pharmaceutical products

Raw Materials

- Pharmaceutical excipients
  (e.g. lactose, magnesium stearate, microcrystalline cellulose etc)

- Solvents
  (e.g. pharmaceutical grade water, methanol etc)

Packaging Materials

- Aluminium foil
- High density polyethylene
- Glass

Pharmaceutical Products

- Tablets
- Injections
- Creams and ointments

Active Drug Substance

- Ranitidine hydrochloride
  (Zantac tablets)
- Salbutamol sulphate
  (Ventolin inhaler)
- Fluticasone propionate
  (Flixonase aqueous nasal spray)

Slides 1-6

Types of Specification

- In-process controls
- Release or manufacturing specification
- End-of-life specification
- Pharmacopoeial specification

Drug Substance Specification

- Identity
- Assay (or potency)
- Drug-related impurities
- Solvent content
- Heavy metals
- Particle size

Drug Substance Properties

- Solubility
- Chemical and physical stability
- Density
- Crystal habit
- Bioavailability
- pKa

Drug Substance Specification

- Assay → 98 to 102% by weight
- Drug-related impurities → not more than 2% total
- Solvent content → not more than 1%
- Heavy metals → not more than 20ppm

Analytical Methodology

- Identity → Infra-red, LC
- Assay → LC, titrimetry
- Impurities → LC, thin-layer
- Solvent content → GC
- Heavy metals → Colour tests, AAS
- Particle size → Light scattering

Analytical Performance

- Typical precision of liquid chromatographic assay methods is around 1% RSD.
- Assay specification is 98 to 102% by weight.

Slides 7-12

Analytical Performance

Key question:

What happens when an assay result of 98.1% is obtained by HPLC?

Does the batch pass or fail specification?

Analytical Performance

Confidence interval is given by:

$$t \frac{s}{\sqrt{n}}$$

Analytical Performance

- Regulatory agencies expect drug-related impurities to be controlled from levels of 0.1% of the active drug content.
- HPLC detection and quantitation limits need to reflect this level of control.

Analytical Performance

Key question:

Does the analytical method (HPLC) have sufficient selectivity and sensitivity to detect and quantify drug-related impurities at or above the level of 0.1% of the active drug content?

Development of the Drug Substance Specification

Batch	Assay (% by weight)
A	98.6
B	98.0
C	101.0
D	100.1
E	99.4
F	99.7

Development of the Drug Substance Specification

Batch	Sodium content (ppm)
A	10
B	3
C	9
D	12
E	2
F	15

Slides 13-18

Development of the Drug Substance Specification

Batch	Water (% by weight)
A	0.22
B	0.41
C	0.88
D	0.69
E	1.02
F	0.14

Development of the Drug Substance Specification

Batch	Methanol (% by weight)
A	0.79
B	0.61
C	3.22
D	0.51
E	0.27
F	0.18

Development of the Drug Substance Specification

Batch	Impurities (% by weight)		
	Principal	Secondary	Total
A	0.9	0.3	1.6
B	0.9	0.3	2.2
C	0.7	0.5	1.5
D	0.3	0.2	0.7
E	0.3	0.2	0.7
F	0.3	0.1	0.5

Review of Batch Data

- The assay data seem unremarkable, although these need to be reviewed in conjunction with total impurity data.
  - mass balance considerations

- A controlled upper limit for sodium may be unnecessary depending on the form in which sodium is present.

Review of Batch Data

- An upper limit for methanol should be set without attempting to encompass Batch C. The result of 3.22% indicates that this batch is atypical and this is supported by other aspects evident in the data review.

- An upper limit for water should be set based on the known chemical (hydrolytic) stability of the molecule.

Review of Batch Data

- The impurity data indicate an improvement in batch quality with time, so that it is likely that batches D, E and F reflect the typical quality of material now achievable during manufacture. Hence, impurity control of not more than 0.5% for the principal, 0.3% for the secondary and 1.0% for the total level is recommended.

Slides 19-24

Summary

- Assessment and assurance of the quality of pharmaceutical materials is established by the use of appropriate specifications for raw materials and products.

- Various types of specifications exist which reflect the need for control during various phases of drug development.

Summary

- Regulatory specifications for drug substances and products require analytical methodology capable of defined levels of performance.

- Final definition of quality specifications for pharmaceutical materials reflect the performance of the manufacturing process, the measurement techniques and the regulatory/safety expectations.

Summary

- Quality review of pharmaceutical materials depends heavily on an assessment against the defined specification and a recognition that the control parameters must be considered collectively rather than individually. In turn, this emphasises the importance of the analytical methodology used to generate the batch data.

Slides 25-27

Concepts of Quality Management and Quality Assurance in Analytical Research Projects and Non Routine Analysis

P. Radvila

Abstract
Contrary to routine analysis, where the sample matrix and concentration range of the analyte are known, the analytical method has been validated and is tried - the method is fit-for-purpose - analysis in non-routine situations and in R&D may be a tortuous problem solving process. Although many basic quality elements are the same as in routine analysis, unknown factors such as unpredictable behaviour of the sample, unknown composition of the matrix and interdependance with the analyte, even uncertainty about the choice of the instrumental method, require not only a new focus on existing quality elements, but - due to the unpredictability of the analytical procedure and extent of effort - also additional organisational quality elements.

The uncertainties in non-routine analysis and R&D require careful consideration of the problem and required or available analytical techniques, planning and organisation of work, supervision of the analytical progress and adequate presentation and quality of results. Project management allows structuring of all aspects - both technical and organisational - of non-routine and R&D work and defines responsibilities and competences of the personnel.
Analytical task quality elements are:
Phase 1: Preparation and planning before starting work
  Definition of task and project design
  Project design and research plan
  Resource management of task
Phase 2: Work in progress
  Progress review/monitoring analysis
The novelty and uncertainties of analytical research and non-routine analysis require additional effort to assure quality. Acceptance of established concepts of quality management and project management is urged.

Slide 1
Non-routine analysis (NRA) and analytical R&D (ARD) are performed regularly in all areas of public concern, e.g. health, environment, in industry and commerce, and especially in academic research. Materials and process innovations as well as novel products, complex combinations of commonly used compounds, traces in uncommon matrices require new or modified analysis methods and motivate the creation of new analytical technologies. Even in routine situations NRA must be performed, when products or processes are „out of routine (control)". Very often the analyst is confronted with samples of unknown origin, doubtful composition

or ill defined analytical request requiring a non-routine approach.

The time and effort devoted to NRA and ARD are – regardless of situations and organizations – much greater than commonly assumed.

Slide 2

Contrary to routine analysis where the complete procedure is defined and straight forward-samples (within a defined range) are known, working methods are defined, documented, e.g. in standard operation procedures as well as proven to give correct results, data reporting is formatted – the uncertainties in NRA and ARD situations usually allow several alternatives and necessitate a search for an optimal analytical work procedure. The capabilities of the laboratory and demands on the results must be met at the same time. High level education and analytical training together with analytical experience and resource-fulness of the personnel are mandatory for efficient execution of work and correct results.

Slide 3

The many unknowns of NRA and ARD require extra effort, from planning of work to the critical review of results. The extent depends on a) the technical complexity and novelty of the analytical task and b) the direct or indirect economic impact of results and risks of incorrect answers. Whereas the direct costs of analytical work can easily be determined, the economic impact of incorrect or delayed results is uncertain, but usually much greater. R&D expenditures in chemistry, in life and health sciences etc. depend very much on costs and speed of analytical developments. Many industrial and environmentally induced investments and international treaties depend on correct data and judicial evaluation of data based on non-routine work.

Slide 4

Basic quality requirements for NRA and ARD are the same as those for routine work.

All customers, external or internal, industrial or academic, expect – the same as for goods or services – that analytical results are „fit for purpose". However, with NRA and ARD not all ultimately attainable quality requirements can be already defined at the outset of work. Fulfilling expectations to a required or self-imposed degree may mean finding a practicable compromise with regard to the initial request and involve parallel planning and execution of work. Quality assurance, e.g. method validation, data verification with reference materials (RM) etc. is an important and integral part of the analytical task. This contrasts with routine analysis where a significant amount of quality assurance (QA) is already performed at the stage of method development.

The ambiguities of NRA and ARD may lead industrial scientists and academic researchers astray by not carefully planning work execution and performance, especially when tests seem trivial.

Slide 5

Due to its characteristics NRA and ARD are problems solving processes. The amount of work – quality and quantity – is not known beforehand. There is a wide range from defined problems to tasks, where the analytical problem has to be determined. All uncertainties endanger straight and efficient progress requiring careful planning and structuring of work as well as control of performance. Activities related to QA, involving both organizational as well as technical elements permeate work at all stages. The amount of time used for QA increases with complexity and novelty of the task. It is safe to say that most organizations, even those without a formal quality system have a set of rules and procedures, which form a loose QA network contributing to success of non-routine work.

Slide 6
QA required for NRA and ARD (above all for all R&D work) is a combination of principles of project management (PM) and elements of quality management (QM). By accepting complex and demanding tasks the analyst (or scientist) assumes to a varying degree the combined responsibilities of a project and laboratory manager. The duties are diverse and time consuming. The approach may differ, due to organizational differences of laboratories in academia, industry, government etc. The responsibilities remain the same.

Numerous textbooks on PM and QM exist. The new quality standards (ISO 9000 series and ISO/IEC Guide 25) and guides (EURACHEM Guidance Document No. 1 "Accreditation for Chemical Laboratories: Guidance on the Interpretation of the EN 45000 series of Standards and ISO/IEC Guide 25") concentrate essential elements of PM and QM based on practical experience. Adherence to all requirements may allow accreditation of a laboratory within the scope of its expertise. Guidance for those performing NRA and ARD and not necessarily seeking accreditation is given in the EURACHEM/CITAC Guide CG 2 „Quality Assurance for Research and Development and Non-routine Analysis".

Slide 7
According to the terms of definition, NRA and ARD work are „tasks" or „projects". The definitions draw a distinction to NRA.

Typical examples where work should be carried out as projects are diploma and doctoral thesis, whereby avoiding deficiencies and undue duration of work. A proposition for teachers and students alike, as well as project managers in general. Novelty and uncertainties require a flexible problem solving process. Analytical work is performed in such a way as to satisfy defined quality requirements. Essential quality elements are presented for judicious selection. They must be integrated and become an inseparable part of analytical work.

Slide 8
NRA and ARD are typically performed in three project phases: initiation and planning, execution and work review, evaluation and reporting of results. Activities in all phases must include those key „quality elements", which assure that work is completed according to the requirements of the customer („Fit for purpose").

The importance of quality elements varies according to the nature and goals of projects and organizations. A wide range of activities related specifically to QA in non-routine situations and research are presented.

Slide 9
At the outset of any given assignment of work the request is analysed, problems and goals clearly defined. Furthermore, it must be determined whether dates can be met, human and financial resources and goals are realistic compared with the

customer's needs and how work and results should be interpreted and exploited. The analyst responsible for the project, e.g. project manager, is advised to clarify all ambiguous points in a dialogue with the customer. Persons or organisations (internal or external) which might be technically involved in the project or are responsible for resources etc. should be consulted as early as possible, if necessary with the consent of the customer.

This stage is more critical than usually assumed; sufficient effort and care should be devoted. A checklist or questionnaire may help to include all pertinent questions and informations.

Slide 10
Determination of requirements and critical review of the analyst's capabilities, e.g. know how and resources, as well as agreement with the customer allows a clear definition and requires documentation of goals, procedures (with alternatives), expense, milestones and dead lines etc. Use of flow or bar charts etc. helps visualize critical areas and points of decision.

Slide 11
Organization of the task environment and resource management are integral parts of project design and documented research plans. Subdividing all activities – especially necessary in complex projects with several methods and numerous collaborators – into self-contained work packages with milestones allows organization of work and its execution, decision making, performance control and communication.

Slide 12
The availability of all needed resources must be taken into account and secured. Agreement with respect to expenditure of work and timing etc. must be sought with all collaborators and persons responsible, especially when auxiliary contributions are requested or when several organizational units are involved.

The amount of planning in NRA and ARD can be considerable, depending on risks and foreseeable difficulties. Replanning in the course of work is very common; planned alternatives allow flexibility and creativity.

Slide 13
Carrying out planned NRA or ARD work in the second phase of the project, the analyst relies to a varying degree on experience, present knowledge, literature data and documented instrumental methods, manuals etc. All activities which may become part of a new analytical procedure, contribute to data as well as project quality, but also deviations and problems must be documented, the main objective of documentation being that work can be repeated and actions are accountable.

Verification and validation of procedures and data are a major part of laboratory work. Use of reference materials (RM) may even involve preparative or synthesis work.

NRA and ARD are only scientifically sound if activities can be duplicated and data reproduced.

Slide 14
Superiors are responsible for monitoring work progress and quality. Customers should be involved when major decisions are made or plans must be modified.

Their review and critique depend on adequate reporting. Reviews include all aspects, laboratory work, results, organizational matters etc. at scheduled acheckpoints and milestones, or irregular intervals, when unexpected problems my require modification of plans. Comprehensive reporting of procedures as well as reasoning of method and decisions during work is part of the results, as much as numerical data. Data should include measurement uncertainty (MU) with description of method validation.

Slide 15
Extent and type of presentation of final results depends on the policies and needs of organizations. Often a sequence of presentations is chosen, e.g. internal report, a standard operation procedure (SOP), diploma or doctoral thesis and external publications. Reports and supplementary statements and commentaries may include an achievement review.

Slide 16
At the final stage the project is reviewed and checked whether all requirements and quality objectives have been met. Often proficiency tests are performed (long) after completion of work; published work may be duplicated and scrutinized for veracity by outside analysts. Final reports, thesis or publications possibly together with archived working documents and samples must enable duplication of results at a later date.

Slide 17
Contrary to routine analysis where quality assurance work is performed beforehand and analytical as well as QA procedures are documented in SOP's, in NRA and ARD quality related activities are performed and documented during the project, as work goes on. Regardless whether a quality system exist or not, NRA and ARD rely on essential key quality elements which must be used in all cases, the extent being adjusted to quality requirements.

Slide 18
NRA and ARD are performed as dynamic processes of work execution and improvement.

QA in NRA and ARD require use of a combination of elements of project and quality management (PM and QM). QA in NRA and ARD is Total Quality Management (TQM). The new standard ISO/IEC 17025 (revision of ISO Guide 25) takes QA for non-routine analysis and research into consideration. Standards and guides are results of training and experience and may help as check lists and guides to good practice also in academic and industrial research where no formal quality systems exist.

Acceptance of quality concepts facilitates the analytical process, ensures efficiency and may assure correctness of results. The analytical education must integrate the use of quality elements in its curriculum. Professional proficiency and flexibility as well as organizational and communicative skills are required.

Typical Non Routine Analysis (NRA) and Research and Development (R&D) Situations

- Question unclear
- Sample and matrix unknown
- Method new or unproven
- Deviation of concentration range

- Synthesis and materials R&D
- Process development
- Analytical R&D (techniques / method)
- "Trouble shooting" (e.g. quality control, production)

- Academia
- Industry and Commerce
- Government and public services

Routine vs NRA and R&D

	RA	NRA /R&D		
Sample	•	•	•	•
Sampling	•	•	•	•
Sample preparation	•	•	•	•
Analysis (method)	•	•	•	•
Confirmation of data	•	•	•	•
Validation / MU*	(•)	•	•	•
Preparation of RM**	(•)	•	•	•
Evaluation of results	(•)	•	•	•
Reporting of results	•	•	•	•
Interpretion / Comment	(•)	•	•	•

* Measurement Uncertainty
** Reference Materials

Characteristics of NRA and R&D

Many unknowns
- Sample, matrix
- Problem, answers
- Methodology
- Resources

Require EXTRA effort (Quality Assurance, QA)
- Planning of task and performance
- Execution and control of work
- Evaluation of work and results

Extent depends on
- Complexity
- Degree of novelty
- Expenditure and duration
- Risks and consequences

Quality Mangement (QM): Quality of Analysis

Requirements
- usefulness of results
- valid and reproducible data
- compliance with cost and time limits

Σ = Fit for purpose

Characteristics and Consequences

- Projects are problem solving tasks
- Many uncertainties
- Flexible planning with alternatives
- Efficiency requires planning and structuring of work
- Much QA effort: technical + organizational

Characteristics and Consequences

Combination of elements

PROJECT MANAGEMENT (PM)
+
QUALITY MANAGEMENT (QM)

Scientist: Manager of the Unknown

Slides 1-6

Definitions (Webster's 1994)

Project

1. Something that is contemplated, devised or planned; plan; scheme
2. A large or major undertaking, esp. one involving considerable money, personnel and equipment
3. A specific task of investigation, esp. in scholarship
4. Educ. an educational assignment necessitating personal initiative on the part of a student

Task

1. A definite piece of work assigned to a person
2. Any piece of work
3. A matter of considerable labor or difficulty

Analytical Task Quality Elements

EURACHEM/CITAC Guide CG2 "Quality Assurance for Research and Development and Non Routine Analysis", Chapters 7.1, 7.2, 7.3

Phase 1: Preparation and planning before work
- definition of task and project design
- project design and research plan
- resource management of task

Phase 2: Work in progress
- progress review / monitoring analysis
- data verification
- changing direction

Phase 3: Work is completed
- achievement review
- reporting, technology transfer and publication
- archiving

---

1. Preparation and Planning
Task Definition + Project Design

Analysis and Definition:

- Understand, analyse, define problem and Review of literature, goals
- Experience
- Sampling, sample / matrix, analyte
- Methods, accuracy, validation, RM
- Dates of preliminary results, completion
- Decision points (date, result, expense)
- Resources, subcontraction
- Confidentiality, intellectual property
- Archiving
- Exploitation

Tool: Questionnaire

---

1. Preparation and Planning
Project Design and Research Plan

Defines

Goals:	Intermediate goals
	Decision points / milestones
	Answers / results
Tactics:	Task
	Workpackage (WP)
Resources:	Personnel
	Equipment
	Total / WP
Time:	Start
	Intermediate dead lines (WP)
	End

Tools:
flow-chart, bar-chart, critical path method

---

1. Preparation and Planning
Task Definition + Project Design

Organization of Task Environment

- Scope
- Objectives
- Resources
- Costing
- Milestone planning
- Contract control
- Financial control
- Change management
- Customer liaison

---

1. Preparation and Planning:
Project Design and Research Plan

Resource Management

- Define authority and responsibility
- Set tasks and time table
- Involve all personnel
- Secure availability (resource, time)
- Monitoring system
- Communicate and consult

Slides 7-12

2. Work in Progress: Record and Verify

Document
- Lab work, observations
- Lab data and print-outs
- Progress, difficulties
- Unsuccessful work
- Deviations from plan
- Ad hoc changes
- Organisational aspects

Verify and validate
- Consistency with literature
- Plausible with experience
- Reproducibility of results
- Check data with RM / model substances
- Reproduction of work

2. Work in Progress: Progress Review

Monitor
- Results, problems
- Deviation from plans
- Expense
- Time schedule vs. achievement

Reports, meetings
- Irregular intervals
- Milestones / dates
- Workpackage
- Major problem

Decide and report
- Change / modify plan
- Stop / go

3. Completion of Work: Reporting

Presentation in reports, SOP, publications:
- Comprehensible
- Enable reproduction of work
- Document trueness (validation, MU)
- Problems, dangers

Achievement review:
- Usefulness of results
- Quality of work
- Gained experience
- Compliance with expense
- Satisfaction of client

3. Completion of Work: Archiving + Reporting

Secure storage:
- Note books
- Data recordings
- Samples
- Reference samples

Requirements:
- Legal / regulatory (contents, length, safety)
- Customer / external agency
- Verification
- Validation
- Proficiency testing with samples
- Duplication of work
- Technology transfer, publication

Key Elements for Quality

- Analyse problem
- Plan and organise work
- Work and monitor results
- Verify and validate data
- Document activities and results
- Report and review
- Archive documents and samples

Combination of PM and QM

Q-Key elements	ISO/IEC 17025 Requirements	Management Tools
Analyse • Plan and Organise • Work and Monitoring	Management	PM
• Verify and validate • Document • Report and Review • Archive	Technical	+ QM

QA in NRA and R&D is TQM !

Slides 13-18

Worked Examples of Teaching Analytical Quality Concepts

Quality Systems for Non-routine and R&D Analytical Work - Accreditation of Non-routine Laboratories

W. Steck

Abstract
These days many testing laboratories are familiar with formal QM-Systems like EN 45001 (ISO Guide 25), ISO 900X (EN 29000 ff) or GLP.

Laboratories engaged in R&D face several QM-systems which are suitable to deal with Non-Routine and R&D analytical work. To make the appropriate choice it is however necessary that the laboratories know the application fields, scopes, benefits and drawbacks of the various systems. Hence this presentation tries to give both survey and insight into the QM-systems concerned like EN 45001, ISO 9001 and GLP as well as the new ISO 17 025. The pros and cons for more or less harmonization of the various standards will also be discussed.

While from the beginning ISO 9001 and GLP allowed more or less efficiently to deal with Non-Routine and R&D the accreditation scheme according to EN 45001 has become flexible enough not until a few years ago. By using the type of test approach to define the scope several accreditation bodies developed a very promising way, which has since then been increasingly practised successfully, e.g. in Switzerland or Germany.

As for laboratories seeking third party recognition of their technical competence accreditation looks more appropriate than the other systems this lecture also addresses some aspects of the accreditation of Non-Routine analysis. Based on own experience the author will give some practical hints on how to prepare a multidisciplinary Non-Routine laboratory according to the type of test approach.

Slide 1
In the first part of the presentation the three quality systems frequently encountered in analytical laboratories (ISO 9001, GLP and ISO Guide 25 respectively EN 45 001) are analysed, compared and assessed with regard to their suitability for non-routine and R&D analysis.

In the second part of the presentation the concept of flexible accreditation of non-routine laboratories on the basis of types of test are introduced. This approach, which is already practised in some European countries, is a successful alternative to the modular combination of existing quality systems, which are dealt with in this lecture, too.

Slides 2 and 3
In the interest of better understanding some definitions at the beginning. Since generally accepted definitions for non-routine and R&D analysis do not yet exist, the ideas on the characteristics of non-routine and R&D laboratories and analysis are summarized.

For the sake of simplicity the terms non-routine and R&D are considered to mean the same thing. The examples used in the presentation stem from the BASF's Central Analytical Laboratory.

The three quality systems are discussed in turn, limiting the discussion to their suitability for non-routine analysis.

Slide 4
EN ISO 9001 is a generic standard containing little detail and designed to be applied almost anywhere. Nevertheless in Chapter 4 (development, design control), it offers the user the possibility of demonstrating R&D capabilities, e.g. to develop products, processes and test methods or to carry out ad-hoc testing reliably.

Slide 5
The results of a study of the literature and assessment of the standard are summarized. In many cases, ISO 9001 can be successfully applied in research and development, especially when the kind of project management now typical of R&D is incorporated into Chapter 4 of the Quality Manual and the associated guidelines. However, various authors do not rate ISO 9001 very highly as a quality system for laboratory-related technical activities.

Slide 6
As an example of how this standard is applied the contents page of a quality management directive used by a BASF production unit concerned with product and process development for chemicals is shown. Design and development guidelines make an important contribution to a systematic and efficient planning and implementation and hence to the successful conclusion of development projects. This is especially true at the key interfaces between research, development, production, marketing and patenting and between some of these activities and environmental, health and safety aspects.

Slide 7
This takes us on to the OECD principles of Good Laboratory Practice, known by its acronym GLP. Among other applications, this system is clearly oriented towards R&D project management.

Slide 8
This is underscored by the numerous R&D capabilities of this quality system listed in slide 8.

Slide 9
However, GLP, which has to be considered to be a legal requirement for registration of (new) chemicals has a number of disadvantages, which outweigh by far the advantages and are listed as cons.

Slide 10
The aspects of non-routine design in EN 45 001 are examined. This orientation of the standard towards testing and calibration and towards the technical competence of the testing laboratory make it a preferred building block of a quality system that is also appropriate for non-routine analysis. Although this standard offers

considerable scope of action and interpretation, especially to technical staff, its flexibility with regard to non-routine analysis could be further improved.

Slide 11
A practical example: Since the BASF Central Analytical Laboratory is mainly concerned with non-routine analysis, appropriate guidelines were incorporated into our quality system on the occasion of our accreditation. The contents page of Chapter 4 of our Quality Manual is shown. In this chapter, for example, the procedure for introducing, modifying or developing of test methods is standardized. Chapter 4 also contains validation and verification guidelines and the statistical methods that are used for this purpose.

In a SOP (standard operation procedure) also guidelines giving details of the procedure for non-routine and ad-hoc testing have been laid down. In these cases some kind of testing programme defines the analytical procedure and validation experiments to ensure valid results. The flowchart contained in EURACHEM/ CITAC Guide 2 is based on a similar diagram in BASF guidelines. Such a flowchart allows systematic planning and implementation of an analytical procedure with a minimum of bureaucracy.

At this stage it is appropriate to make a brief provisional appraisal. From what has been discussed so far it can be summarized that in its current form none of the three standards mentioned can be recommended as more suitable than the others. Each system has a different focus and potential and both pros and cons with regard to its application in non-routine analysis.

Slide 12
This is illustrated in the following table, which compares some characteristics important for non-routine analysis. The aim of the table is to highlight the potential of each of the three systems. The assessment is based on an evaluation by a group of analysts at BASF laboratory. It is immediately apparent from the table that modular combinations of these systems would yield more suitable quality systems. Such a combination should in all cases include EN 45 001 because of its emphasis on technical competence, as well as elements from ISO 9001 and/or GLP.

Slide 13
This approach is illustrated again in slide 13. The area in the broken line is intended to indicate that EN 45 001 is also considered alone, following appropriate flexibilization, to be suitable for non-routine analysis as will be shown in the second part of this talk.

Slide 14
This overhead shows two examples of modular combinations, which should now no longer come as a surprise. The merging of ISO 25 / EN 45 001 with ISO 9001 to form the new ISO 17 025 also makes this new standard more suitable for the accreditation of non-routine analysis, as was intended.

Less well known is the combination of EN 45 001 and the Dutch national standard NEN 3417 employed in the Netherlands for the accreditation of R&D analysis. NEN 3417 is used as a supplement to EN 45 001 and comprises only the "good" aspects of GLP.

This is where EURACHEM/CITAC Guide 2 also comes into its own. Designed to be applied to non-routine analysis these guidelines too are essentially based on a modular concept involving technical competence, administrative elements and R&D project management.

The first part of the talk can be summarized by the statements that there are already a number of quality systems that can be successfully applied to non-routine analysis, whether with or without formal accreditation or certification. From a global point of view, some of the solutions presented can be viewed as regional approaches only. International organizations such as EURACHEM and CITAC are concerned with bringing about a worldwide harmonization. An example is the EURACHEM/CITAC Guide 2 presented here. ILAC (International Laboratory Accreditation Cooperation) is another organization that is increasingly involved with the subject of accrediting R&D laboratories according to globally standardized criteria.

Slide 15

The second part of the presentation focuses on the accreditation of non-routine laboratories according to EN 45 001. It is shown that assessing competence on the basis types of test instead of just test methods makes accreditation considerably more flexible. A number of routine and non-routine laboratories in countries such as Germany have already been accredited in this more flexible way, including the BASF Central Chemical Laboratory.

Slides 16 and 17

With these two overheads the meaning of the terms "testing field" and "types of test" is explained.

Slide 18

Accreditation based on types of test offers laboratories further advantages, for example with regard to including new test methods within the scope of an already accredited type of test. However, as far as the test items examined by the laboratory are concerned, the accreditation scope must continue to be described in detail.

Slide 19

Compared with conventional accreditation based entirely on test methods, one of the novel aspects of accreditation based on types of test is that a selection of proofs of technical competence is assessed, so that the scope of the laboratory's competence in a certain field need be confirmed by a limited number of examples only. The assessment is based mainly on this exemplary selection and no longer on all of the laboratory's test methods.

Besides validated test methods, the following are also recognized as proofs of competence for the accreditation of non-routine analytical laboratories: generic test methods, standard procedures relating to types of test, standard procedures for non-routine analysis, scientific reports, and documents resulting from any reasonably sophisticated laboratory work (test reports, test certificates, etc.)

Slide 20
The basic conditions for selecting exemplary proofs of competence is illustrated in order to get the approval to a certain type of test, taking as example unlimited type-of-test accreditation of gas chromatography. The documents to be selected for GC must reflect adequate competence, e.g. concerning the technical range as kind of injection, separation, detection and application (test items).

Slide 21
Summary of the presentation.

Slide 22
A list of relevant literature, including the outstanding publications of Eggimann and Morkowski from EMPA, Switzerland, dealing with the concept of the modular assessment of the competence is provided.

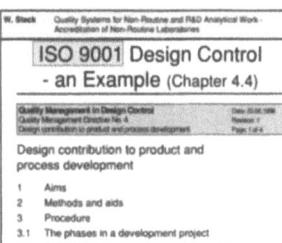

Slides 1-6

The OECD Principles of GLP (Paris1998) Research Design

- GLP is a quality system concerned with the organisational process and the conditions under which studies are planned, performed, monitored, recorded, archived and reported.

- One purpose of GLP is to promote the development of comparable quality test data.

The OECD Principles of GLP (Paris1998) Research Design

Pros
- Study Director
- Study Plan
- Knowledgeable personnel in sufficient number
- Adequate and controlled test equipment with
  - - where appropriate - traceable calibration
  - Selected Standard Operating Procedures (SOP's)*
  - Validated Computerized Systems
  - Evaluation and discussion of results
  - Trackability and Repeatability (Reports, raw data, archive)
- Test Facility Management
- Facilites
- Quality Assurance Programme

*when designed as generic test methods

The OECD Principles of GLP (Paris1998) Research Design

Cons
- Restricted flexibility, too high degree of details
- Formality superior to proper performance
- Highly bureaucratic
- Excessive Control
- Continuing tendency to excessive requirements
- Expensive/Unfavourable cost-benefit-ratio
- Not adapted to the needs of chemical analysis

EN 45001 / ISO Guide 25 Research Design

Pros
- Focusing on technical competence
- (Potential for more) Flexibility
- Adapted to analytical testing
- Higher scope for self-responsibility
- Generic test methods applicable

Cons
- Continuing tendency to expand requirements
- Professional judgement excluded from accreditation scope*

*The new standard ISO/IEC 17025 allows the inclusion of opinions and interpretations in the test report

Development and Research
(Chapter 4 of ISO 9001)

Contents

4.1	Aims and application range
4.2	Definitions
4.3	Responsibilities
4.4	Procedures
4.4.1	Development of test methods
4.4.2	Modification of test methods
4.4.3	Non-Routine testing with respect to accreditation
4.4.4	Validation of test methods
4.4.5	Use of statistical methods for the validation of test methods
4.5	Documentation and archiving
4.5.1	Documentation
4.5.2	Archiving

Comparison of QM-Standards for Non-Routine and R&D Analytical Work

Field	ISO 9001	EN 45001/ ISO 25	GLP
Organisational Process Project Management Study Director, Study Plan	+		++
Technical Competence	+(+)	++	+
Staff	+(+)	+(+)	+(+)
Equipment	+(+)	++	+(+)
Facilities	+	++	+(+)
Evaluation and Discussion of Results/Professional Judgement			++
Quality Assurance Programme Audits, Reviews	++	+(+)	+(+)

Slides 7-12

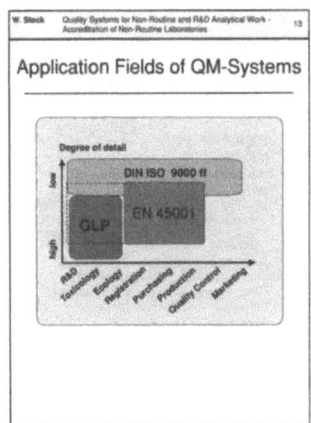
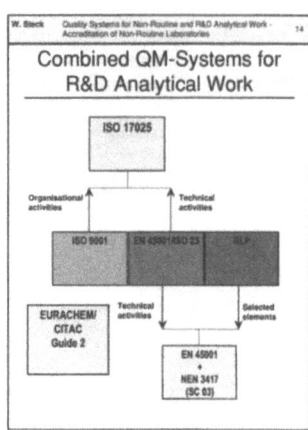

Slides 13-18

Accreditation by Types of Test

Proofs of competence:

Competent technical staff

Knowledge and experience concerning measurement principles, methodology, technology, technical range, science, test items and validation

Documented competence like specific test methods, generic test methods, SOPs for non-routine analysis, SOPs related to types of test, scientific reports, test reports and others

Adequate and controlled/calibrated equipment

Accreditation of Type of Tests Strategy for Selection of Test Methods for GC

Technique			Evaluation method	Application field analyte/matrix (working range)
Sample application	Separation	Detection		
split injection	capillary column	flame emission	internal standard	purity control for release of raw materials and final products (> 90 g/100 g)
splitless injection	packed column	thermal conductivity	external standard	
cold on-column injection	column coupling	mass selective	standard addition	impurities in chemical products (0.01 – 10 g/100 g)
temperature programmed vaporization		element specific	area %	residual monomers e.g. VC in PVC (µg/g – g/g)
Headspace GC		nitrogen specific	isotopic dilution	
Pyrolysis GC		thermal energy analyser		N-nitrosamines in air (pg/m³)
Thermo desorption		electron capture		Enantiomer separation

QM-Systems for Non-Routine and R&D Analytical Work - Accreditation of Non-Routine Laboratories

Summary

I. In its current, non-modified version no single QM-System can be recommended as beeing the most suitable for monitoring the quality of Non-Routine and R&D Analytical work.

II. Combining the technical focus of EN 45001/ISO 25 with organisational and R&D project management elements taken e.g. from GLP or ISO 9001 has proven as fit for purpose.

III. Using the types of test approach the accreditation according to EN 45001/ISO 25 has become flexible enough to deal not only with Routine but also successfully with Non-Routine and R&D Analysis.

IV. Non-Routine / R&D Laboratories have already been accredited in Europe according to the schemes II. or III.

Accreditation of Non-Routine Laboratories

Publications concerning flexible accreditation

1. EAL-P 10, Edition 1, February 1997, The Scope of Accreditation and Consideration of Methods and Criteria for the Assessment of the Scope of Testing.
2. DACH GmbH, Technische Mitteilung, No. 3, Februar 1997, Flexibilisation of the Scope of Accreditation, Accreditation of Type of Tests
3. F. Eggimann, J. Morkowski: An adaptable modular QA-system of a large multi-discipline testing laboratory. Proc. of the 1st Eurolab Symposium, January 28 to 30th, 1992, p. 24-34.
4. F. Eggimann, J. Morkowski: A modular approach for the assessment of the competence of a testing laboratory in a specific testing field. Proc. of the 1st Eurolab Symposium, January 28-30, 1992, p. 501-515.
5. F. Eggimann, R. Fischbach, J. Thiers: Experiences with the modular assessment of conformity and competence of a testing laboratory, 2nd Eurolab Symposium, Testing for the years 2000, Florence, Italy, April 94.
6. EURACHEM/CITAC Guide 2, 1998, Annex C
7. Steinhorst, Accred Qual Assur (1998) 3: 294

Slides 19-22

Evaluation of Uncertainty in Analytical Measurement

M. Rösslein, B. Wampfler

Abstract
The following contribution deals with the contents of two lessons of 45 minutes each. Basic knowledge of the evaluation of measurement uncertainty in analytical chemistry is imparted. After the two lessons the audience should know the new concept according to GUM/EURACHEM [1,2] in its fundamentals and be able to calculate the measurement uncertainty of a simple analytical procedure. The content is structured into three parts:

The first part starts with a short description of random and systematic influences. Based on this information the old but still applied concept of uncertainty evaluation is presented. That concept distinguishes between the measurement uncertainties "typeA = random errors", and "type B = systematic errors". Then the teacher leads over to the new concept according to GUM/EURACHEM. It is indicated that the new concept allows the transformation of nonstatistically evaluated uncertainties into standard uncertainties. Thus the calculation of the combined standard uncertainty is possible.

In the second paragraph it is shown how to calculate the combined standard uncertainty. In addition the required theoretical knowledge is interposed. The common procedure is demonstrated by means of the corresponding flow chart of the EURACHEM Guide. Each step is illustrated separately. For the identification and the analysis of the uncertainty sources the cause and effect diagram from Ellison and Barwick is initiated. The triangular- and rectangular distribution is presented without deriving the formulas of the variance. The comments about the calculation of the combined standard uncertainty refer only to independent variables. They show the law of propagation of uncertainty and derive two simple mathematical rules for addition/subtraction and multiplication/division. The case of correlated variables is only mentioned without treating. Finally, it is indicated how the results and their measurement uncertainties are put on record correctly.

The lectures require only basic mathematical know-how from the students. It is a qualified instrument for the introduction of the measurement uncertainty at universities and technical colleges. However, it has to be taken into consideration that only an introduction can be given in a double lesson and exercises of at least eight hours should follow.

Slide 1
When carrying out a series of measurements, due to randomly fluctuating influence quantities (temperature, humidity, etc.) the individual values are not identical which means that the measured results are scattered around a mean value. This run to run variation of results is very often a normal distribution or an almost normal distribution in analytical chemistry.

Slide 2
Let us assume that n independent observations q_k were carried out under the same experimental conditions but which were subjected to randomly fluctuating influence quantities. How can the values of the measurand q be estimated? In most cases the arithmetical mean value $\bar{q}$ provides the best estimate. It is obtained by the first equation. $\bar{q}$ is the mean value of n observations of the measurand q and q_k is an individual value of an observation of the measurand q. The experimental variance $v(q_k)$ of the observations is given by the second equation. The variance $v(q_k)$ and its square root, the experimental standard deviation, $s(q_k)$ characterise the variability of the observed value q_k or more specifically the distribution around the mean value $\bar{q}$.

Slide 3
In addition to the variance of individual values, there is also a variance $v(\bar{q})$ of the mean value. This variance of the mean value $\bar{q}$ is the quantity of the uncertainty of the mean value $\bar{q}$. $s(\bar{q})$ is the standard deviation of the mean value $\bar{q}$. The larger the number n of independent observations is the more $s(\bar{q})$ becomes narrow and therefore the better the estimation of the mean value $\bar{q}$ is. On the other hand $s(q_k)$ is not dependent on n but the estimation of $s(q_k)$ becomes better if n is increased.

Slide 4
In addition to the randomly fluctuating influence quantities, which cause a frequency distribution of the results, systematic effects also have to be taken into account with every measurement made. A systematic effect causes a shift of the randomly distributed individual values with their mean value with respect to the true value. The resulting systematic deviation can be seen as the difference between the mean value and the true value. When, for instance, a pH measurement is carried out, a temperature difference between the calibration solution and sample solution causes a systematic deviation of the measured value to the actual pH value. Systematic deviations are not dependent on the number of measurements carried out and therefore their values cannot be reduced by increasing the number of measurements. The result of a measurement should always be corrected with respect to known systematic effects. Nevertheless, it must be borne in mind that these corrections are subject to uncertainty, too.

It is important not to confuse randomly fluctuating or systematic influence quantities with spurious errors or measurement deviations. Spurious errors (blunder) are caused by human error or are due to an incorrect function of a machine. Transfer errors such as reversing figures when noting the results of a measurement or air bubbles in the cell of a photospectrometer are typical examples of spurious errors. Measurements where a spurious error is discovered should be totally ignored and not included in evaluation statistics. Spurious errors are not included in the calculation of measurement uncertainty.

Slide 5
We differentiate between random and systematic measurement deviations. The term „error", often used in this connection, is confusing and should not be used

any more. Measurement uncertainty describes a range or distribution of possible values. For instance 82 ± 5 describes a range of values. For this reason measurement uncertainty differs from the expression „error", which is defined as an individual value. This is the difference between a result and the true value. Up till now it has been usual to designate random deviations as error type A and systematic deviations that have not been corrected as error type B.

Very often the error estimate is only determined for the random or statistically detectable measurement deviation in the form of the standard deviation from a series of measurements. Therefore the contribution of several steps in a procedure to the random deviation of the complete procedure is not taken into account. Systematic deviations are also seldom corrected or they are combined to a great extent in the unsatisfactory form of type B error. Thus very often the measurement uncertainties quoted are far too low. This gives rise to discussions on the correctness and therefore on the validity of the results of an analysis because the results from different analyses cannot be directly compared with each other. A correct decision is not possible until the combined uncertainty is known which includes random uncertainty and the uncertainties of systematic correction. This combined uncertainty cannot be determined within the old definition.

Slide 6

The difficulties experienced with the definition of measurement uncertainty used till now resulted in 1980 in two fundamental recommendations being issued by the Comité International des Poids et Mesures (CIPM):
- The prime aim of any efforts must be to achieve the comparability of results and the unproblematic further processing of uncertainties quoted.
- This aim can only be achieved by a system change when defining measurement uncertainty: uncertainties determined using statistical methods on the basis of repeated measurements will be designated as type A estimation of uncertainties. All other uncertainties that cannot be calculated by a statistical procedure will be designated as type B estimation of uncertainties.

EURACHEM has ascertained that this definition of uncertainty has created some confusion with the old designation (type A ⇔ random, type B ⇔ systematic). Therefore EURACHEM recommends that the design-ations A and B be dropped and instead to use the expressions statistically calculated uncertainties and uncertainties determined by other means.

Slide 7

This slide illustrates a new concept prepared at an international level on the basis of recommendations from the CIPM. The measurement results should be corrected for all identified systematic deviations and the uncertainties (remaining deviations) of the corrections combined with the random measurement deviations to an overall measurement uncertainty.

In principle, measurement uncertainty also includes all unknown systematic deviations. However, because not even the order of magnitude of these measurement deviations are known, they are not taken into account in the calculation of the measurement uncertainty. When an analytical method has been validated, meaning that it has been approved as fit for a given purpose, an unknown systematic deviation should not occur or at least have an insignificant impact on the combined measurement uncertainty.

In order to determine the remaining deviation both standard deviations, produced by a test series for determining the correction of a systematic deviation, and also estimates taken from minimum or maximum values are included in the calculation of measurement uncertainty. This fact means that it is possible to efficiently implement the new concept for calculating measurement uncertainty into daily work in the laboratory.

Slide 8
The new concept is based on observable quantities. It leads to a combined standard uncertainty of the corrected mean value. This uncertainty is clearly larger than the standard deviation of a mean value that is defined under repeatability conditions and that has not been corrected with respect to systematic effects.

Slide 9
Slide 9 shows the new definition of measurement uncertainty. Using the concept described here, the calculation of the measurement uncertainty is decoupled from the true value. This decoupling from the true value offers several benefits. According to this definition, the true value is a value which corresponds with the definition of an observed special measurement quantity. The true value is not known and it can only be determined with the aid of a perfect measurement. This means of course that the true value can never be determined.

When the earlier method was used to calculate measurement error great efforts were made to determine the true value. This is a paradox in itself because by definition it is impossible. In contrast the new concept demands that the measurement procedure should be fully considered and scentinized so that all influence quantities are determined and recorded. When all influence quantities are known and corrected for, the results from different analyses or analytical procedures will correspond. This agreement is possible without having to know the true value.

Slide 10
In most cases the best estimate of a measurand's value, which is subject to randomly fluctuating influence quantities, is the arithmetic mean value. The standard uncertainty of the mean value is defined from the standard deviation. The mean value and the standard uncertainty are calculated according to slide 3. If the random fluctuations from the observation of the measurement quantity are correlated, for instance linked with one another over time, then the mean value $\bar{q}$ and the standard deviation of the mean value

$s(\bar{q})$ are not suitable for correctly describing the measurement results. In such cases the task must be limited to observing individual values as a function of time.

Slide 11
The standard uncertainty $u(x_i)$ of influence quantities, whose value is not or cannot be defined by repeat measurements, is determined by making an estimate which is based on various pieces of information about possible effects. Possible information sources are, for example

- previous measurement data
- experience with or knowledge of the behaviour and characteristics of the object under investigation and of the measurement technique being used
- information quoted by the manufacturer
- data based on calibrations or certificates
- uncertainties quoted from reference data taken from manuals.

The optimum use of all available data for non-statistical estimation of standard uncertainty demands wide knowledge and experience on the relationships within a measurement. The necessary skills can be learnt by practical application of non-statistical estimating of uncertainties.

It is important to point out that non-statistical estimates of uncertainty are just as good or even better than standard uncertainty determined by experimental means, especially when the experimental standard uncertainty has been calculated from very few measurement values (see slide 12).

Slide 12

The fact that measurement uncertainty has been determined using statistical or non-statistical methods does not permit any conclusions to be made on the quality of the values obtained. This will depend to a much greater extent on the experience and skills of the person who collected the data for determining the measurement uncertainty. Uncertainties determined statistically are also subject to uncertainty (uncertainty of uncertainty). The more individual measurements are carried out, the more certain the standard uncertainty. This is illustrated clearly by this slide which lists the uncertainty of the uncertainty for a number of measurements.

For purely statistical reasons, the uncertainty of the standard uncertainty (standard deviation) for the mean value q can be surprisingly large with a limited number of random samples. For $n = 10$ observations it is still 24%. The figures listed in the table show clearly that the standard uncertainty of the standard uncertainty for a number of random samples ($2 < n < 10$), which is common in practice, cannot be ignored. Therefore standard uncertainty estimated on the basis of tolerance limits of a complete series of measurements and an assumed frequency distribution (triangular distribution), is often better than the standard uncertainty determined using, for instance, three measured values.

Slide 13

The calculation of measurement uncertainty can be divided up into the following steps:
- Specification of the measurand

  The specification of the measurand describes what is to be measured and how.
- Identification of the uncertainty sources

  A list of possible sources of measurement deviations is drawn up for each single operation. When looking for possible uncertainty sources, chemical reactions during the analysis must be taken into account and also any assumptions in this connection.
- Quantification of the uncertainty components

  The degree (value) of uncertainty of every identified source of measurement deviation is determined either statistically or estimated using other means.

- In order to permit the calculation of the combined standard uncertainty (u_c) carried out in the next step according to the law of uncertainty (error) propagation, every uncertainty component must be converted into standard uncertainty (u_i).
- Calculation of the combined standard uncertainty (u_c)

All uncertainty components are combined into the combined standard uncertainty according to the law of uncertainty propagation (slide 23).
- Do the relevant components have to be re-evaluated?

After having calculated the combined standard uncertainty (u_c) as described in the previous step, the contribution of every individual uncertainty source to the total uncertainty can be calculated. Then the evaluation of the standard uncertainty of the relevant contributory factors should be studied critically. In addition the uncertainty sources which contribute very little to the combined uncertainty can be eliminated from further processing steps. This procedure permits the calculation of measurement uncertainty to be considerably simplified.

Independent of the calculation of the measurement uncertainty, it is advisable to use checking samples of certified reference material in series measurements and to ensure the statistical monitoring of a measurement procedure or of a test instrument (e.g. using control charts).

Slide 14
Write down a clear statement of what is being measured, including the relationship between the measurand and the parameters (e.g. measured quantities, constants, calibration standards etc.) upon which it depends. Where possible, include corrections for known systematic effects. The specification information, if it exists, is normally given in the relevant Standard Operating Procedure (SOP) or other method description.

Slide 15
It is recommended to draw up a list containing all sources of uncertainty which are relevant for an analysis method. This procedure should be structured in order to prevent influence quantities being forgotten or counted several times. To produce this structure it is advantageous to use a cause and effect diagram recommended by Ellison and Barwick [3]. It can be constructed employing four steps:
- Write the complete equation for the result. The parameters in the equation form the main branches of the diagram.

Slide 16
- Consider each step of the procedure and add any further factors to the diagram, working outwards from the main effects.
- For each branch, add contributory factors until effects become sufficiently remote, that is, until effects on the result are negligible.

Slide 17
- Resolve duplications and re-arrange to clarify contributions and group related causes.

Duplicated influence quantities are produced from the detailed adding of effects for every individual parameter. This can be demonstrated in the following two examples: The value of a particular influence quantity varies from measurement to measurement. These variations are contributory factors to the repeatability of the complete procedure. When the repeatability of the complete procedure has been determined experimentally and included in the measurement uncertainty, no additional recount of variations in the values of the individual influence quantities is performed.

Example: Often all weighings in a procedure are carried out using the same analytical balances. If the uncertainty of the calibration for each weighing is taken fully into account, it could be overestimated. These considerations produce the following rules for simplifying the cause and effect diagram.

Cancelling effects: remove both. For example, in a weight by difference, two weights are determined, both subject to the balance "zero bias". The zero bias will cancel out of the weight by difference, and can be removed from the branches.

Similar effect, concruent in time: combine into a single input. For example, run-to-run variation on many inputs can be combined into an overall run-to-run precision "branch". Some caution is required; specifically, variability in operations carried out individually for every determination can be combined, whereas variability in operations carried out on complete batches (such as instrument calibration) will only be observable in between-batch measures of precision.

Different instances: re-label. It is common to find similarly named effects which actually refer to different instances of similar measurements. These must be clearly distinguished before proceeding.

Slide 18

Uncertainties that have been determined statistically are usually available as experimental standard deviations and can be used directly as standard uncertainty. If the information is available in the form of a confidence interval, the standard uncertainty (standard deviation) must be calculated before any further processing by dividing it by the corresponding expansion factor.

When determining uncertainty using non-statistical methods, the following procedures are recommended (slides 20 - 24).

For a great deal of data that is available for calculating measurement uncertainty, only the upper and lower limits of the measurement quantity are known, rather than the distribution of the individual values in the interval. Let us take as example the volume of volumetric pipettes in the quality class A. The manufacturer usually quotes tolerances without any additional information. Using this data it is not possible to determine the type of frequency distribution of the actual volumes. However, we can assume that the manufacturer only delivers volumetric pipettes whose actual volume is within the quoted ranges. In addition the manufacturer controls the production process in such a way that the possibility for a certain volume is greater near the specified value than near to the tolerance limits.

Slide 19

A triangular distribution is a reasonable way of describing this frequency distribution. a_- is the lower tolerance limit and a_+ is the upper one.

Slide 20
The expected value $\bar{q}$ for the volume is the middle point of the interval between the two tolerance limits with the associated variance given by the second formula.

If the difference between the tolerance limits $a_+ - a_-$ is designated as $2a$ the expression for the variance $u^2(q)$ can be simplified. The standard uncertainty is derived from the square root of the variance.

Slide 21
On the assumption that the value q_i of the measurement quantity occurs more regularly in the middle of the interval than near the limits a_- and a_+, a triangular distribution was chosen. In some individual cases the assumption that the frequency remains the same within the total interval is more realistic. In this case the triangular distribution described in slides 19 and 20 is replaced by a rectangular distribution.

Slide 22
The expected value for the measurement quantity is the middle point of the interval between the two tolerance limits a_- and a_+. The associated variance is given by the second equation. The standard uncertainty is again derived from the square root of the variance.

Slide 23
We assume that the influence quantities are independent on each other. The combined standard uncertainty $u_c(y)$ is the square root of the combined variance $u_c^2(y)$, which is calculated with the help of the given equation. This equation is based on an approximation of the first term of a Taylor series $y = f(x_1, x_2, ..., x_N)$ and expresses what is designated as the uncertainty (previously: error) propagation law. The partial differential $\partial f / \partial x_i$, often called the sensitivity coefficient, describes how the result y is varied by a small change in the influence quantities $x_1, x_2, ..., x_N$. For many applications, that equation can be reduced to a simplified form.

Slide 24
For equations containing only additions and subtractions of influence quantities, e.g. $y = k(p + q - r + ...)$ with k as constant, the combined standard uncertainty $u_c(y)$ is given by the relationship in the first frame. For equations containing only multiplications and divisions, e.g. $y = k(p q ...)$ with k as constant, the combined standard uncertainty $u_c(y)$ is given by the formula in the second frame.

Slide 25
Rules 1 and 2 will be implemented in practice with the first equation as example. First of all we substitute $(o + p)$ by z and $(q + r)$ by n. The standard uncertainty is calculated for z and n according to Rule 1.

Slide 26
Afterwards the intermediate result is combined according to Rule 2.

Evaluation of Uncertainty

There are often cases in which the combined standard uncertainty cannot be determined for given functional relationships with the aid of rules *1*) and 2). For these cases we have to fall back on equation on slide 23. This equation is only applicable if the influence quantities x_i are independent of each other or uncorrelated. If at least two x_i are significantly correlated this fact must be taken into account.

This can be demonstrated by a chromatography example. Considering the method of the internal standard the signal of the internal standard is correlated with the one of the sample. As both signals contribute as a ratio to the equation of the measurand, not the uncertainty of the individual component should be estimated but the uncertainty of the ratio should be evaluated. In this way both correlated parameters are reduced to one uncorrelated parameter. Correlated influence quantities are not dealt with here any further.

The standard uncertainty u_c can be transferred to the expanded uncertainty U by multiplication using an expansion factor of 2 or more. If sufficiently certain data is available the expanded uncertainty can be assigned approximately to the confidence limit. If the expanded uncertainty is used in the results, the expansion factor must always be quoted.

Slide 27

If $u_c(y)$ is used as the quantity for the uncertainty, the numerical measurement result should be preferably quoted as follows:

m = 100.02147g with (a combined standard uncertainty) u_c = 0.35mg.
Alternatively they can also be quoted by the following method, however, a definition is necessary in order to prevent confusion.

m = 100.02147(35)g, in which case the figures in brackets give the value of u_c (combined standard uncertainty) which refers to the last digits of the measurement result quoted.

m = 100.02147(0.00035)g, in which case the figures in brackets give the value of u_c (combined standard uncertainty) in the same unit as the measurement result.

m = (100.02147±0.00035)g, in which case the figures after the ### sign is the figure value of u_c (combined standard uncertainty) and is not a confidence range.

The last format with the ### sign should be avoided, if possible, because traditionally this is used to quote a range indicating a high level of confidence and this could cause confusion with an extended measurement uncertainty. The text parts in brackets can be left out if they have been defined previously.

Slide 28

What is the measurement uncertainty under given experimental conditions, if the contents of a sodium hydroxide solution (NaOH) is determined using the titrimetric standard potassium hydrogen phthalate (KHP)? We now intend to use the previously introduced concept to answer this question.

Slide 29

Procedure: the contents of a freshly prepared CO_2-free NaOH solution is determined volumetrically using the titrimetric standard KHP under a protective gas. For this approx. 0.5g of the titrimetric standard is weighed, water is added and the mixture stirred until the standard is fully dissolved and homogeneously distributed. The standard solution is titrated against the NaOH solution.

Slide 30
The equation of the measurand is given in this slide. The concentration of sodium hydroxide is calculated by multiplication of the initial weight of potassium hydrogen phthalate with the purity of the titrimetric standard KHP. This product is divided by the consumption of caustic soda solution and the molecular mass of KHP. Because we need the concentration of sodium hydroxide in mole per liter, but the consumption of sodium hydroxide solution is given in milliliter, the calculated quotient is multiplied by 1000.

Slide 31
The four parameters provide the following initial situation for setting up the cause and effect diagram. The measurand is the concentration of the *NaOH* solution c_{NaOH}; it is assigned to the horizontal main arrow. The four influence quantities from the previous equation form the four main branches.

Slide 32
The main branches are supplemented by other influence quantities until all potential factors relevant for measurement uncertainty have been collected.

Weight of the titrimetric standard: The calibration and the repeatability of the weighing provide contributions to the measurement uncertainty. The uncertainty of the calibration is produced to a great extent by the deviation from the linear behaviour, by the intercept and by the sensitivity drift. The intercept cancels and can be removed because the final weight is a weight by difference. If the repeatability is determined during several autocalibration periods of the analytical balance, the sensitivity drift is part of the repeatability.

Volume of the NaOH solution consumed: Its uncertainty is affected by the calibration of the piston burette, the repeatability of the dosage, the expansion of the titer by the impact of temperature and finally by the uncertainty of the end-point detection. With regard to the impact of temperature, it is to be noted that the expansion of the glass of a piston burette is negligible compared with that of the solution. When considering the end-point detection, the repeatability and the bias from the determination of the equivalent-point contribute to the uncertainty.

Purity and molecular weight of the KHP: The other main branches of the diagram, the purity of the KHP and its molecular weight are not supplemented by more branches. Thus we have the present diagram.

Slide 33
Normally when validating an analytical procedure the repeatability of the complete procedure is determined. This is also assumed in this example and therefore the repeatability of the individual operations are combined in one branch. This means that the data already available from the validation can be used.

Slide 34
The resulting diagram looks as follows.

Slide 35
The influence quantities listed in the cause and effect diagram will now be

quantified. The weighed sample of KHP is 0.511g. In order to estimate the uncertainty caused by the non-linear characteristics of the balance, we consult the manufacturer's user manual for the instrument. There we discover that the non-linearity of the balance is limited to ± 0.15mg. So we already have the upper and lower limit. Because we do not have any information for a possible distribution of the values within the quoted tolerances, we transfer the interval into the standard uncertainty assuming a rectangular distribution.

Slide 36

Here we first of all turn to the end-point determination. During validation the titration procedure was optimized in a manner that there is no bias due to inhomogeneity of the solution or due to the response time of the pH electrode. In addition transfer reactions of hydrogene ions are very fast in aquatic solution. For this reason there is no need to correct the consumption of NaOH solution and therefore there is no uncertainty for a bias.

With a weight of 0.51g KHP the consumption of NaOH solution is approximately 25 mL. According to the manufacturer's specification, the calibration of the 50 mL piston burette is within the tolerance of ± 0.05 mL. From this, assuming a triangular distribution, the standard uncertainty of the burette's calibration is calculated.

Slide 37

During a year the temperature in the laboratory deviates not more than approx. ± 4°C from the calibration temperature of the burette.. The uncertainty of the measured volume based on this variation can be calculated according to $u(V_{Temp}) = V \cdot \alpha \cdot u(T)$. α is the expansion coefficient for water which is $2.1 \cdot 10^{-4} \, °C^{-1}$. Assuming a triangular distribution we have a standard uncertainty of 0.009ml.

Slide 38

Because the contributions to the uncertainty of the consumption of NaOH solution behave additively, the standard uncertainty is produced from the square root of the sum of the square of both components (rule 1). The standard uncertainty for the consumption of the NaOH solution up to the end-point is 0.022 mL.

Slide 39

According to the manufacturer's information, the contents of the titrimetric standard, with reference to the theoretical amount of acid to be expected, are within the range of 99.86% to 100.14%. The factor for purity is therefore 1.000 ± 0.0014. We can assume that the purity was determined by analytical chemical means and that the uncertainty quoted is based mainly on the random scatter of the measurement values. Therefore we calculate the standard uncertainty for purity assuming a triangular distribution.

Slide 40

The titrimetric standard KHP has a sum formula $C_8H_5O_4K$. The standard uncertainty of the molecular mass of the compound is calculated from the uncertainty of the atomic mass of the elements of the molecule. A table of the atomic weights with their tolerances is published every two years in the "Journal of Pure and Applied Chemistry".

Slide 41

The standard uncertainty for each element is calculated from the tolerances published by IUPAC assuming a triangular distribution for the frequency distribution. Table on slide 41 indicates the individual contributions of the elements to the molecular mass together with their uncertainty contributions. The uncertainty for the molecular mass of several of the same atoms is calculated by multiplying the standard uncertainty of the atomic mass of one element by the number of atoms. The relative weight of KHP is

$$F_{KHP} = 96.088 + 5.0397 + 63.9976 + 39.0983 = 204.2236 \text{ g/mol}.$$

Slide 42

Because this expression is the sum of independent values, the standard uncertainty $u(F_{KHP})$ is calculated by extracting the square root from the sum of the quadratic contributions (Rule 1).

F_{KHP} is derived from the sum of the individual element contributions. Because these are the sums of individual atomic contributions, it can be assumed that, because of the law of propagation of uncertainty, the uncertainty from each element contribution is calculated from the sum of the squares of the individual atomic contributions. For the example with carbon this would give

$$u(c) = \sqrt{8 \cdot 0.00041^2} = 0.00016 \text{ instead of } 8 \cdot 0.00041 = 0.0033.$$

For this the following should be noted: Rule 1 "Sum of the squares" only applies for contributions which are *independent* of each other. In this particular case an element contribution is obtained by the multiplication of a published value and not by the sum of results of a related multiple determinations. The uncertainty of the element contributions is therefore derived from the corresponding multiplication of the standard uncertainty of the atomic contribution. In contrast the contributions of the various elements are independent of each other and can therefore be dealt with using the normal rule of uncertainty propagation.

Slide 43

When carrying out validation, the repeatability of 0.1% was determined for the complete analytical procedure. This value is available in the form of a standard deviation which is interpreted as standard uncertainty in the calculation of measurement uncertainty.

It is important that at regular intervals, the repeatability is again determined for the complete analytical procedure. This value must agree with the value from the validation within the quoted tolerances. Only under these conditions can the value for repeatability, determined during validation, be used in the calculation of measurement uncertainty without further consideration. In addition, the data gained by the continuous determination of the repeatability ensures that a measurement procedure is kept under statistical control. This is also one of the most important preconditions for guaranteeing the quality of the analytical data on a long-term basis.

Slide 44

The standard uncertainty of the parameters for determining the *NaOH* concentration and the repeatability of the complete procedure are listed in the table of this slide.

Slide 45
The concentration of caustic soda is calculated.

Slide 46
The combined standard uncertainty of the concentration of caustic soda is calculated with the help of this relation (Rule 2).

Slide 47
When the values listed in the slides 44 and 45 are inserted in the previous equation (slide 46) the calculation result in a combined standard uncertainty of $1.5 \cdot 10^{-4}\,\text{mol}\cdot\text{L}^{-1}$.

Slide 48
In order to be able to compare the contributions of the individual uncertainties with the combined standard uncertainty, in the following histogram displays the square of the relative uncertainties in percent.

From this we can see that the largest contributions to uncertainty are produced by the volumes of the NaOH solution and from the repeatability of the complete procedure. The contributions from the weighings and the relative molecular weights are negligible. If the uncertainty of the procedure had to be improved, the first to be optimised would be the volumes of the NaOH solution. Possible measures would be to carry out titration in an air conditioned room and to use a high quality piston burette.

References
[1] Guide to the Expression of Uncertainty in Measurement
First edition (1995) ISBN 92-67-10188-9
International Organization for Standardization
[2] EURACHEM Guide "Quantifying Uncertainty in Analytical Measurement"
First edition (1995) ISBN 0-948926-08-2
[3] S.L.R. Ellison, V.J. Barwick, Accred Qual Assur (1998) 3: 101-105

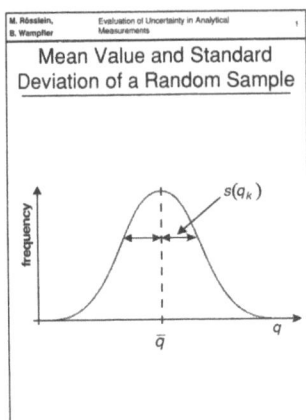
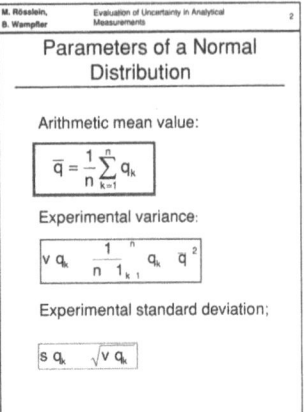
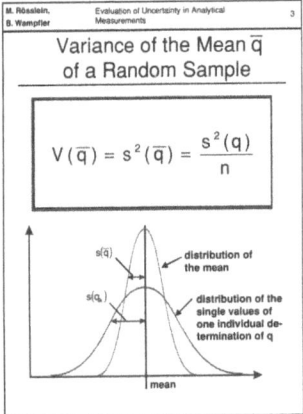
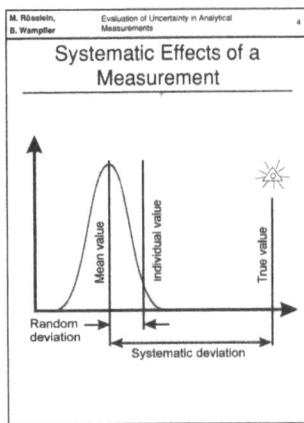
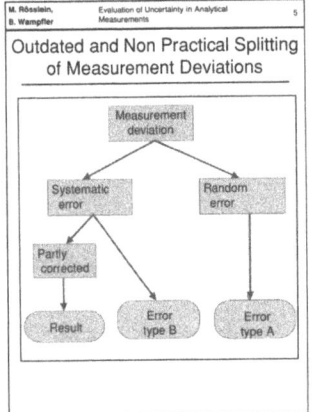
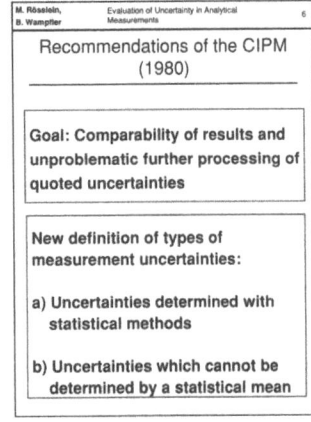

Slides 1-6

Slides 57

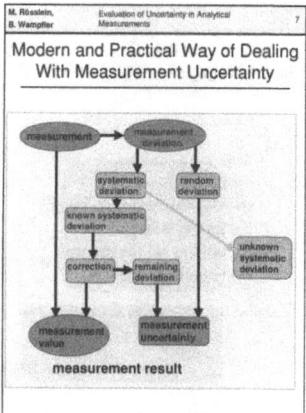

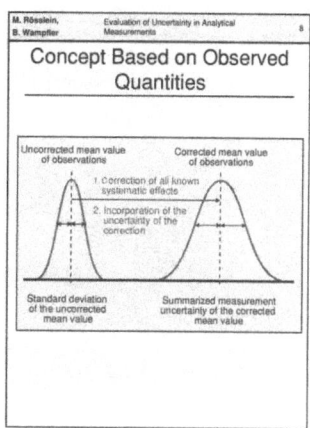

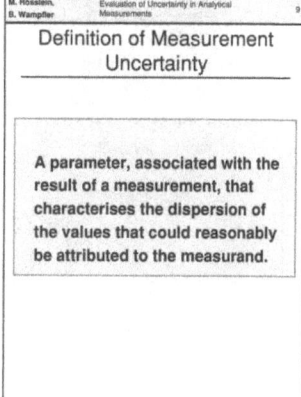

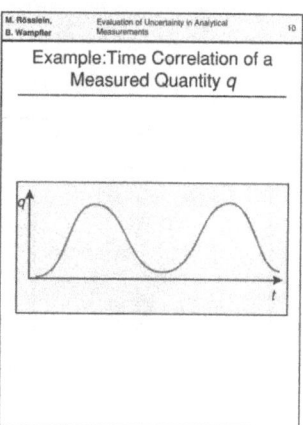

Determining Measurement Uncertainty Non-Statistically

Possible sources of information:

- previous measurement data
- experience with the sample and the measurement technique being used
- information quoted by the manufacturer
- data based on calibrations or certificates
- uncertainties taken from manuals

Uncertainty of the Experimental Standard Uncertainty

Numbers of measurements n	Uncertainty of the Uncertainty / %
2	76
3	52
4	42
5	36
10	24
20	16
30	13
50	10

Slides 7-12

58   M. Rösslein, B. Wampfler

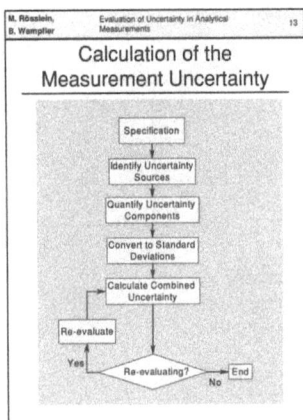

Calculation of the Measurement Uncertainty

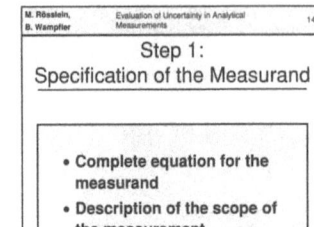

Step 1: Specification of the Measurand

- Complete equation for the measurand
- Description of the scope of the measurement
- Correction for the known systematic effects

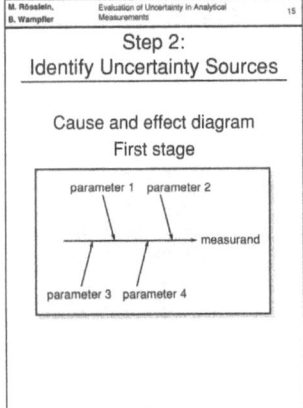

Step 2: Identify Uncertainty Sources

Cause and effect diagram
First stage

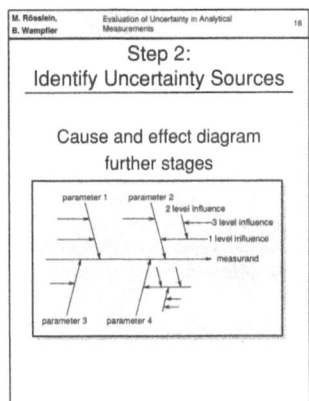

Step 2: Identify Uncertainty Sources

Cause and effect diagram
further stages

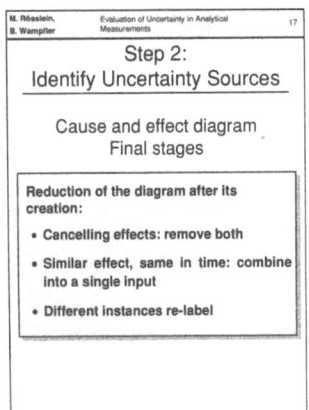

Step 2: Identify Uncertainty Sources

Cause and effect diagram
Final stages

Reduction of the diagram after its creation:
- Cancelling effects: remove both
- Similar effect, same in time: combine into a single input
- Different instances re-label

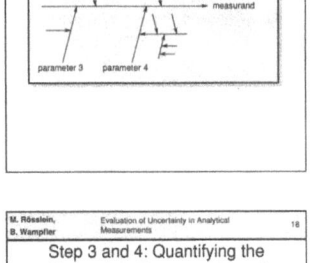

Step 3 and 4: Quantifying the Uncertainty Components and Conversion into Standard Uncertainty

Example: Usual tolerances for some volumetric pipettes

content [mL]	colour - code	tolerance [mL]
1	blue	0.007
5	white	0.015
10	red	0.020
25	blue	0.030
50	red	0.050
100	yellow	0.080

waiting time 15s

Slides 13-18

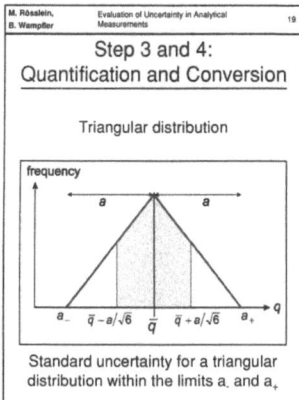
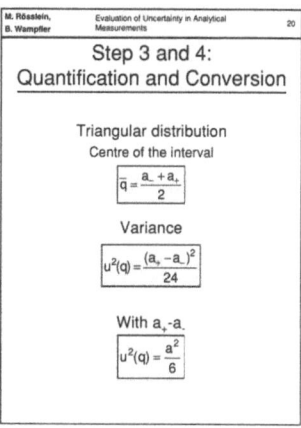
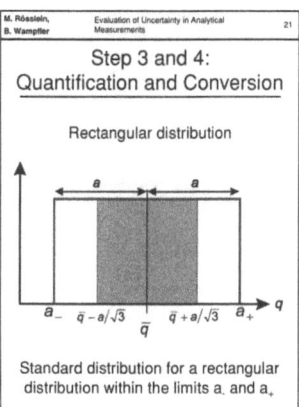
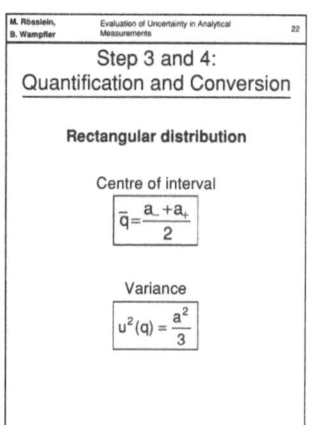
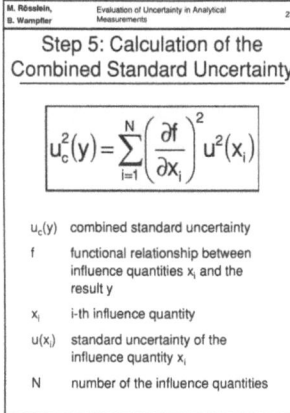
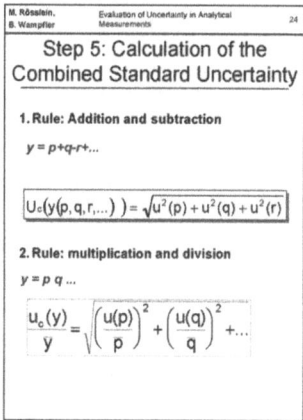

Slides 19-24

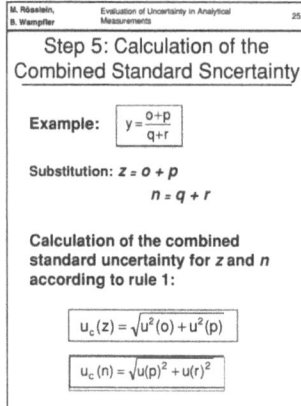
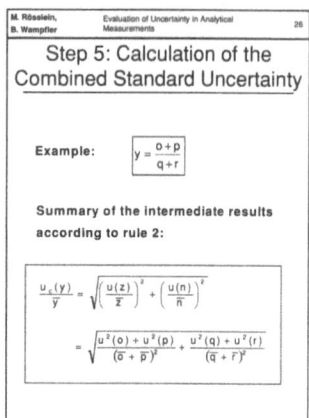
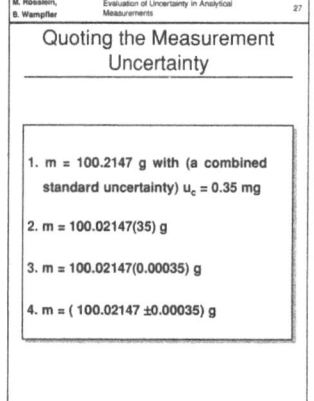
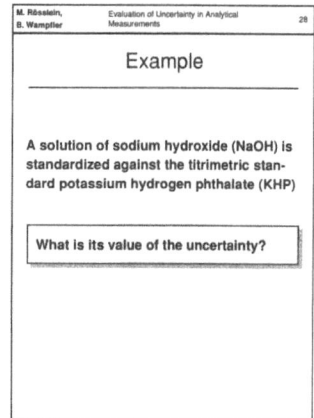
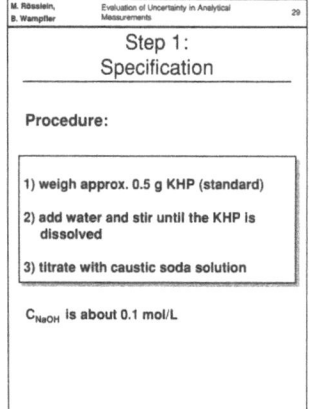
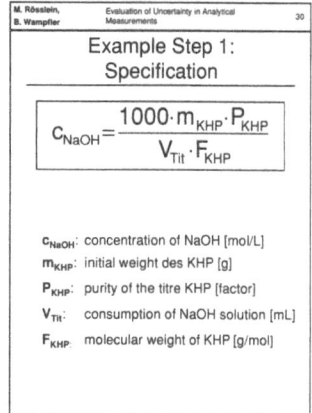

Slides 25-30

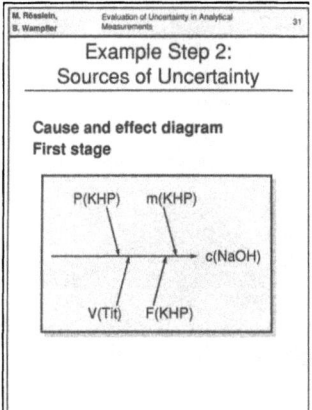
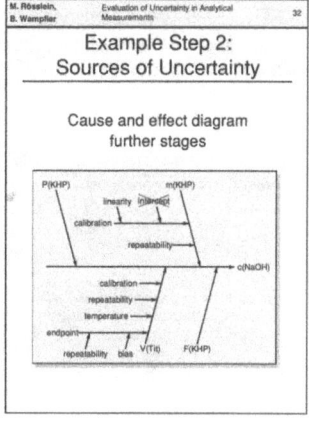
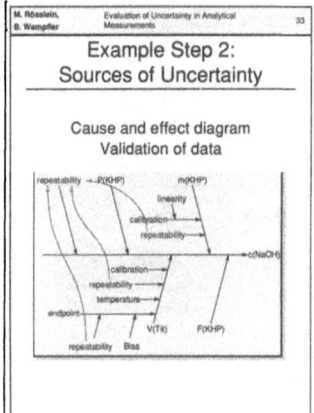
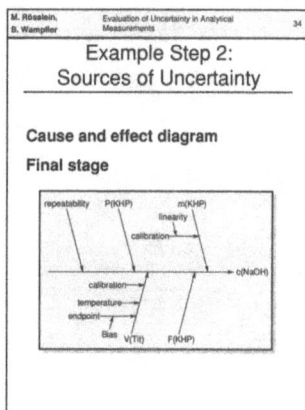
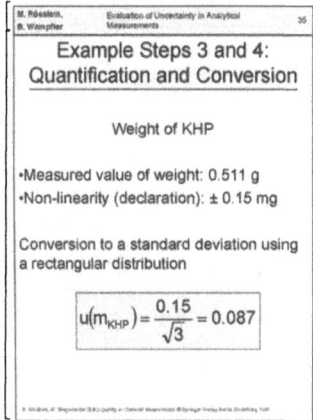
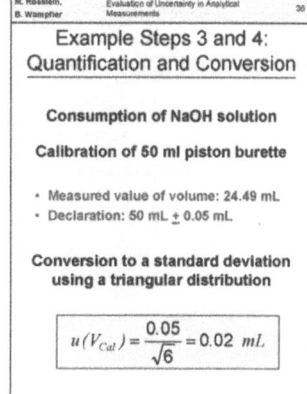

Slides 31-36

Example Steps 3 and 4: Quantification and Conversion

Consumption of NaOH solution

Expansion of the NaOH solution as a result of temperature variation

- Variation of temperature : ± 4°C
- Expansion coefficient of water: $2.1 \cdot 10^{-4}$ °C^{-1}

Conversion to a standard deviation using a triangular distribution

$$u(V_{Temp}) = \frac{25 \cdot 2.1 \cdot 10^{-4} \cdot 4}{\sqrt{6}} = 0.009 \; mL$$

Example Steps 3 and 4: Quantification and Conversion

Consumption of NaOH solution

Standard uncertainty

$$u(v_{Tit}) = \sqrt{u^2(v_{Cal}) + u^2(v_{Temp})}$$
$$= \sqrt{0.02^2 + 0.009^2} = 0.02 mL$$

Example Steps 3 and 4: Quantification and Conversion

Purity of the standard

- Declaration: 99.87% - 100.14%
- Factor: 1.000 ± 0.0014

Conversion to a standard deviation using a triangular distribution

$$u(P_{KHP}) = \frac{0.0014}{\sqrt{6}} = 5.7 \cdot 10^{-4}$$

Example Steps 3 and 4: Quantification and Conversion

Molecular weight of KHP
sum formula: $C_8H_5O_4K$

element	atomic weight	published tolerance
C	12.011	± 0.001
H	1.00794	± 0.0007
O	15.9994	± 0.0003
K	39.0983	± 0.0001

Example Steps 3 and 4: Quantification and Conversion

Molecular weight of KHP
Standard uncertainty

Assumption: triangular distribution

element	molecular weight	standard uncertainty single element	standard uncertainty all elements
C_8	96.088	0.00041	0.0033
H_5	5.0397	0.00029	0.0015
O_4	63.9976	0.00012	0.00049
K	39.0983	0.000041	0.000041

$F_{KHP} = 204.2236$

Example Steps 3 and 4: Quantification and Conversion

Molecular weight of KHP
Standard uncertainty of F_{KHP}

$$u(F_{KHP}) = \sqrt{(3.3 \cdot 10^{-3})^2 + (1.5 \cdot 10^{-3})^2 + (4.8 \cdot 10^{-4})^2 + (4.1 \cdot 10^{-5})^2}$$
$$= 0.0037$$

Slides 37-42

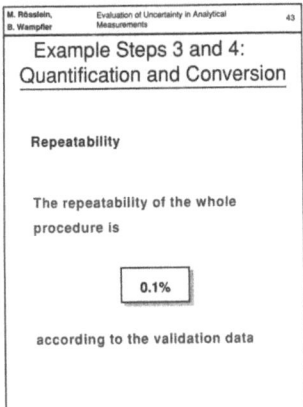
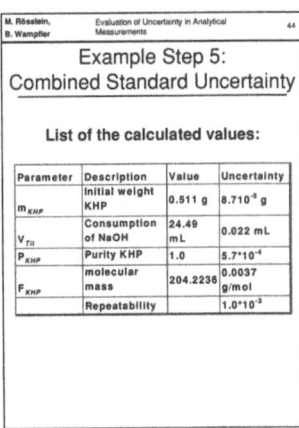
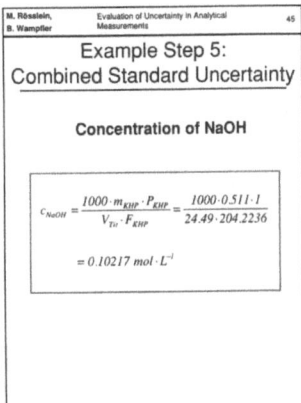
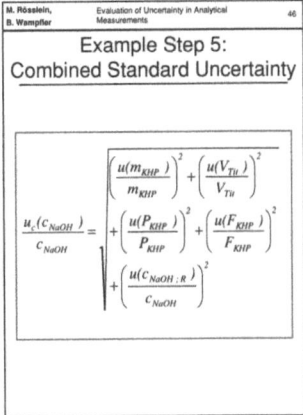
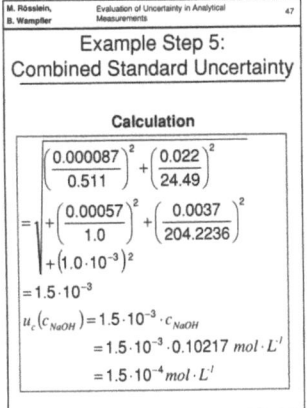
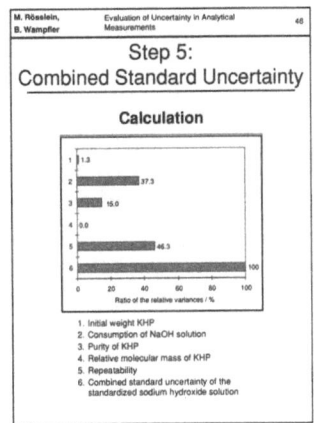

Slides 43-48

Traceability / Trackability

A. Ríos

Abstract
Traceability is a key metrological concept which is defined by the International Vocabulary of Basic and General Terms in Metrology as „the property of a result of a measurement or the value of a standard whereby it can be related to stated references, usually national or international, through an unbroken chain of comparisons all having stated uncertainties" [1]. From this definition, it is clearly stated that any experimental measure is based on a comparison with references and that the knowledge of the uncertainties of the values to be compared is absolutely necessary. Therefore, traceability allows the comparability and the harmonisation of results between analytical laboratories. Hence, demonstration of traceability is the primary objective in chemical measurements.

Recently, a wider meaning of the traceability concept has been recognized. In fact, there is an etymological meaning related to the history of the generation of a product or the behaviour of a system. Thus, the ISO 8402-94 („Quality Management and Quality Assurance Vocabulary") defines traceability as „the ability to trace the history, application or location of an entity by means of recorded identifications". This additional facet adds to a richer concept of traceability. Some authors also proposed complementary terms, like for example *trackability* [2, 3]. They reported the following definition: „the property of a result of a measurement whereby the result can be unique related to the sample". It means that the result of a measurement can be linked unambiguously to the sample to which it refers.

All these tracing connotations will be presented and discussing their implications from the quality point of view of the results provided by the laboratories, as well as the quality of the activities involved in the production of analytical reports.

Slide 1
Within the 2nd EURACHEM Workshop on Education & Training, a summary of the main aspects concerning with *traceability/trackability* terms (and the corresponding meanings) was presented. The following main points will be addressed:
- Traceability as the foundation of metrology.
- Definitions and meanings of traceability/trackability terms.
- The third point is an open question: do we really need the 'trackability' term?
- At the end some final remarks will be pointed out.

Slide 2
Traceability is the foundation of any metrological science. Particularly, in chemical analysis, Analytical Chemistry deals with the way to perform and

implement the metrological principles in chemical measurements: Analytical Chemistry is therefore the science of chemical measurements, and hence, traceability is one of the intrinsic foundations of Analytical Chemistry. In fact, that traceability means it has always been present in the work carried out in analytical laboratories for years. Only lately this term has been increasingly used in the Analytical Chemistry domain because of the development and world-wide implementation of Analytical Quality Systems.

Having a look to the definitions given in the dictionaries, three different words can be found:
- Traceability (noun): "the quality of being traceable"
- Traceability (adjective): "capable of being traced' and 'suitable or of a kind to be attributed"
- Traceability (verb): "related to carrying out a study or to the history of something."

Slide 3

The basic idea involved in metrology is the association of *results* with the corresponding *system under study* through the implementation of one or more measurement(s). This simple principle can be easily implemented (in general) in physical measurements, in which -commonly- the measurement process is mainly performed by the use of previously calibrated equipment. On the contrary, in the case of chemical measurements, the analytical process connecting samples with results, involves several steps. Only the measurement step can be calibrated, but not the rest of steps. Thus, the overall analytical process must be *validated*, and this activity is longer and more complicated than a simple calibration. In this case, appropriate validation ensures the traceability of the results.

Slide 4

However, what about the official definitions and, specially, the meanings of these important metrological terms?

Slide 5

Regarding the official definitions given by the ISO, it is interesting to check the evolution in the traceability definition. It is a good example that metrology (and hence, the metrological principles and definitions) is nothing finished or definitely established. This is a positive aspect because it demonstrates that it is a living science, continuously adapting to the new needs.
Thus, the definition given by the ISO in 1984 stated:

The property of a result of a measurement whereby it can be related to appropriate standards, generally international or national standards, through an unbroken chain of comparisons.

Its graphical meaning can be seen in the figure: analytical results are, basically, expressed as a concentration (a relationship between the mass of analyte and the weight of sample); the chain of comparisons must reach the standard kilogram.

Now, the present definition given in 1993 [1], incorporates three important aspects highlighted in this slide:
- It also includes the value of a standard,
- it admits stated references for comparisons, and
- it recognises the importance of the uncertainties to make the comparisons.

Slide 6
According to this definition, measurement standards play a key role. Therefore, a clear hierarchical classification of standards according to their traceability is necessary. This classification of standards shows the traceability chains and distinguishes between the S.I. basic standards, the chemical standards, and the analytical standards [4].

Slide 7
The role of measurement standards is decisive, in order to achieve comparability of results between laboratories. Through primary traceability chains (A) laboratories meet traceability requirements with respect to basic (S.I.) standards by using chemical standards. If, as expected, traceability between chemical standards is assured, traceability/comparability between laboratories is achieved (secondary traceability chains, B).

Slide 8
Another aspect directly derived from the ISO definition is the fact that an analytical result must be expressed by two values: the first one (M) is the best estimation of the measured parameter, and the second one (N) the doubt in such estimation. Analytical (or statistical) and metrological terms are derived from M and N. It is important to distinguish the similarities and differences between both groups of terms; specially in this case, between *traceability* and *accuracy*. Accuracy of analytical results implies their traceability, and - at the same time - the objective of traceability is to achieve the accuracy of results. However, traceability replaces the true statistical value by the reference value (a real value instead of a hypothetical value).

Slide 9
Because the *analytical process* can be considered as a „*black-box*" linking samples with results, the traceability of the results produced will be basically assured through:
- The proper traceability of the measurement standards used.
- The *calibration* of all the equipments involved in such process.
- The *validation* of the overall analytical process.
- In fact, it can be stated that *validation* has to meet the other two requirements.

Slide 10
During the validation of the analytical process, traceability of the analytical results must be demonstrated; for instance by comparison against the certified value of a CRM (certified reference material) with a similar matrix to the sample. The use of a „t" statistical test is the way to do that.

Slide 11
When an appropriate CRM is not available, other references can be used, as it is shown in this slide [5].

Slide 12
Some illustrative examples about the traceability chains involved in different types of analysis are shown:

- A gravimetric determination.
- A titrimetric determination, and
- An instrumental method of analysis.

These examples can be very useful for teaching the concept of traceability to students during their Analytical Chemistry courses.

Thus, in a primary method of analysis (traditionally called „absolute method"), as for instance the gravimetric determination of iron, two experimental data must be obtained: the weight of the sample and the weight of the calcinated product (if 100% pure). If the basic traceability chain involved in the calibration of the analytical balance is ensured, then traceability of the results can be achieved. In the calculation, two others chemical standards are involved: the mole and the ^{12}C.

Slide 13

In the case of a titrimetric determination, additional links should be included in the traceability chain. Thus, a calibrated burette is needed, as well as a primary standard with demonstrated traceability. This standard can be directly used as titrant solution or can be used to standardise the solution of a working standard used to perform the titration of the sample solution.

Slide 14

In the case of the comparative methods of analysis by using analytical instruments, the situation is more complicated to assure the traceability of the results. Two main weak points can be identified in the general scheme in which instrumental analysis is based. The first one is common to the other two examples previously shown: the sample preparation necessary to perform the analytical measurement. However, the second one is characteristic of this type of analysis: the calibration of the instrument involved in the analytical measurement. Specially, the so called *"analytical calibration"* whereby the relationship between the instrumental signal and concentration of the analyte must be defined. This is a decisive activity to assure traceability of the results obtained in instrumental analysis, and hence the established model must be appropriately validated.

Slide 15

On the other hand, the ISO, in the *quality management and quality assurance vocabulary* defines *traceability* as the "Ability to trace the history, application or location of an entity by means of recorded identifications" [6]. It represents another facet of the traceability term dealing with a *tracing connotation* very close to the definitions given by the language dictionaries. Thus, from this point of view traceability may have different meanings:

- In a productive sense, its relationship to the origin, history, distribution and location.
- In a calibration sense, the relationship to standards, constants, properties or reference materials.
- In a data collection sense, it relates calculations and data with sample results.

Slide 16

This additional facet is quite coherent with the fact that the objective of traceability is to achieve the accuracy of the results (the centre of the blank in the figure) but this objective can be reached by following different ways defined by the chain of comparisons.

Slide 17
It seems clear that both definitions must be taken into account:
- That reported by the ISO Vocabulary of Terms in Metrology, and
- That appeared in the ISO QM/QA Vocabulary.

Slide 18
Therefore, to consider an *integral approach* of *traceability* the two main facets recognised by the ISO should be mixed: the first one related to the history of the generation of a product or the behaviour of a system; and the second one involving the relationship to references. Hence, why not adopting an integral definition including both? Something like that proposed at the bottom of this slide.

Slide19
This integral approach to traceability involves not only the relationship to measurement standards through tangible objects (v.i., a CRM), but also with not very clear tangible analytical tools (methods or procedures); and even exclusively written standards, mainly represented by national and international *norms*.

Slide 20
In 1996, J. Fleming, B. Neidhart, Ch. Tausch and W. Wegscheider proposed a new term named *trackability* [2, 3]. These authors defined this term as *"the property of a result of a measurement whereby the result can be uniquely related to the sample"*. It implies that every step of an analytical method has to be documented in a way that can be linked unambiguously to the sample to which it refers. Therefore, it describes traceability to a sample. Thus, all samples must be uniquely labelled; all operations performed on a sample must be recorded in a notebook or computer system; chromatograms, spectra and other instrumental outputs must be labelled with the sample identification; etc.

Slide 21

Afterwards, these authors accepted that trackability „is not a property of the result of a measurement but rather a property of the system", and they recognised that it is possible to state a corrected definition [3]:

"The property of a system which enables the ready retrieval of the different elements of a record to allow unambiguous correlation with a uniquely identified sample".

Slide 22
In fact, this new term is a matter of discussion. I agree with the authors that the new terms should be generated following the „bottom up" approach, trying to find a broad consensus among the users, and then ask ISO to adopt the new term. However, at the moment, I think we are just in the consensus step. In my opinion, the main question is:

Do we really need this new term or is it better to expand the traceability definition?

It is without doubt that the meaning given by the authors to this term is absolutely necessary and basic under a quality system program.

Slide 23
With respect to the question posed above, Valcárcel and Ríos published in 1995 a general article on traceability [4]. In this article a wider meaning of the term traceability was presented, in which the „traceability of samples" was also incorporated. In fact, as it is shown in the slide, the samples that are finally subjected to the analytical process feature two concatenated traceability connotations that materialise in a unbroken, cyclic chain of relationships. On the one hand, the samples should be representative of the economic and/or social problem addressed, which materialises in an analytical problem that requires definition of the object; this is represented by a set of bulk samples from which the aliquots finally subjected to the analytical process are selected [7]. On the other hand, the samples that are actually subjected to the analytical process and the results they produce should also be unequivocally related. In the first case, representativeness of the results and to sample integrity are referring; whereas in the second case the so-called „sample custody chain" is involved. By assuring the horizontal and vertical chains, coherence between results and social/economic problem exists, and hence traceability of such results.

It is clear that the trackability meaning is implicit in this scheme: the unambiguous link between the object to be studied and the results.

Slide 24
Perhaps a wider definition of traceability may be thus proposed which incorporates the trackability meaning:

Property of a result of a measurement, linked unambiguously to the sample to which it refers, or the value of a standard whereby it can be related to stated references, usually national or international, through an unbroken chain of comparisons all having stated uncertainties; and the ability to trace the history, application or location of all the elements involved in the generation of such result by means of recorded identifications.

Slide 25
According to this proposed (integrated) definition, and as is reported in a recent article [8], by this extended approach, it is possible to speak about traceability of a result, a standard, an equipment, a sample and a method.

Slide 26
Thus, the traceability of a result has both facets (related to its sample and to standards), but moreover it refers to the documentation of the generation of such result. It implies human, material, instrumental, methodological, environmental and temporal documentation about its generation.

Slide 27
The traceability of a CRM, not only deals with the certified values, but also with the documentation on its preparation. It involves the origin of the material (natural or artificial), if the analytes have been spiked on the material or they were originally contained in it, the studies performed on both homogeneity and sta-

bility, etc. In this respect, there is an interesting example reported by P. Sandra [9] about the determination of PCBs in soil samples by extracting these compounds by SFE (super fluid extraction). When a SRM (secondary reference material) in which the analytes were spiked to a clean soil was used, the recoveries obtained under different experimental conditions were very high (almost 100%), but when a CRM in which the analytes were originally contained was used, the best recovery was about 30%. It seems clear that it is important to know the way in which the certified material has been prepared, because the results obtained will be different depending on a set of factors (for instance, the origin of the analytes, as in the example).

Slide 28
The traceability of an analytical equipment is also a recognised term that involves several aspects. Thus, a detailed and updated record containing information about the installation, calibrations and corrections, hours of use, samples processed, etc., is necessary. This is also consistent with the ISO 8402-94 definition given before. According to the norms ISO 9000 or EN 45000, this kind of information is an essential part of the Quality Manual, and is also an element of the Good Laboratory Practices.

Slide 29
Traceability of analytical methods is a key aspect within metrology in chemistry [10]. First of all, it is important to again state the decisive role played by the internal validation of analytical methods. This is an absolutely necessary activity, because „official", „standard", „reference" or other methods externally validated do not assure the quality of the results. Today, the so-called *"traceable method"* is a term used with preference. It refers to a method that produce results (with their uncertainties) characterised by traceability to established references (CRMs or primary methods). From this it becomes clear which role CRMs play in chemical measurements. The high metrological quality is indeed claimed for the primary measurement method, having particular characteristics: its operations must be thoroughly known and explained; its uncertainty must be thoroughly described in terms of basic SI units; and its results must be accepted without reference to standards of the measured quantity. These requirements can only be found in a few analytical techniques, such as gravimetry, coulometry and isotope dilution-MS (the traditionally called absolute methods of analysis).

Slide 30
Some main points can be stated as final remarks. Thus, with respect to the *traceability* term/concept, two general approaches can be considered. The first one connects with the orthodox view of traceability, expressed by the official positions in both QM and metrological domains. Within the QM domain, it refers to the characteristics of an entity or/and measuring equipment. Within the metrological definition, it refers to the property of a result or the value of a standard. However, by the incorporation of a (quasi-)heterodox approach, the traceability concept is enriched and it covers basic aspects within the quality system programs. On the one hand, it deals with the metrological characterisation in the chemical measurement domain, applicable to aspects such as results, standards, equipments,

samples and methodologies. On the other hand, it refers to tracing aspects documented in quality assurance systems.

With respect to the term *trackability*, I think the priority is to give answer to a couple of questions: Do we accept this new term or we admit that it is an implicit facet of the concept of traceability? On the other hand, do we need an expanded definition of traceability? A forum of discussion is necessary in order to clarify that we can defend from a scientific-technical point of view, and that we can teach our students.

Slide 31

At the same time, some challenges should be pointed out within the traceability/trackability domains.

First, as a consequence of this discussion forum, it is a basic objective to make these concepts understandable and accessible at the analytical bench level, so that they are no longer the exclusive patrimony of theoreticians and bureaucrats.

Thus, the concept of traceability will be more flexible and practical.

Within this practical side, to admit with clarity that traceability to well established standards is also valid, practical and useful.

To avoid useless arguments. For instance, to avoid pointless discussions such as whether to use the mole instead of the kilogram as the ultimate reference in chemical metrology.

And, finally, with respect to our responsibility, to systematically introduce these concepts in Analytical Chemistry education and research.

References

[1] International Vocabulary of Basic and General Terms in Metrology.
 International Organization for Standardization (ISO), 1993.
[2] J. Fleming, B. Neidhart, Ch. Tausch and W. Wegscheider.
 Accred Qual Assur, 1 (1996) 43.
[3] J. Fleming, H. Albus, B. Neidhart and W. Wegscheider.
 Accred Qual Assur, 1 (1996) 234.
[4] M. Valcárcel and A. Ríos. Analyst, 120 (1995) 2291.
[5] X. Rius and J. Riu. SMT Workshop on Metrology. Madrid, 1998.
[6] Quality Management and Quality Assurance Vocabulary. International
 Organization for Standardization, ISO 8402-94.
[7] A. Ríos and M. Valcárcel. Analyst, 119 (1994) 109.
[8] M. Valcárcel and A. Ríos. Fresenius J. Anal. Chem., 359 (1997) 473.
[9] P. Sandra, private communications
[10] P. de Bièvre, Accred Qual Assur, 5 (2000) 171

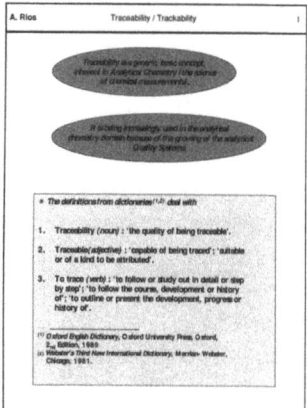
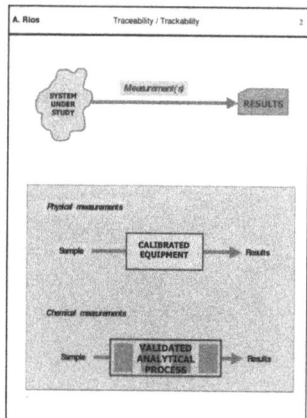
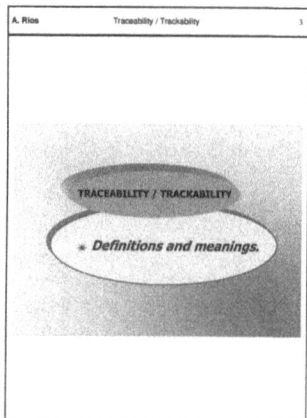
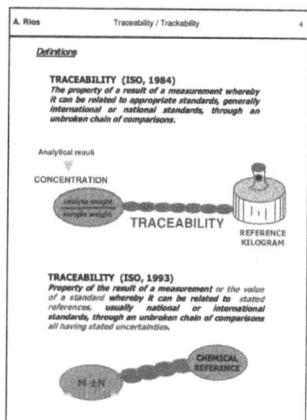
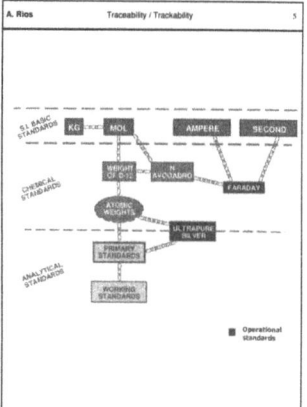
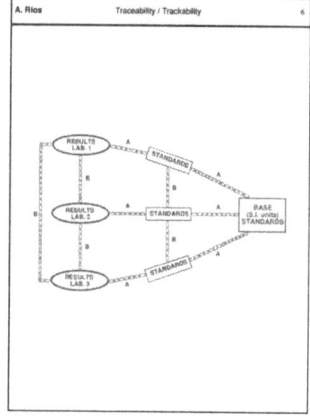

Slides 1-6

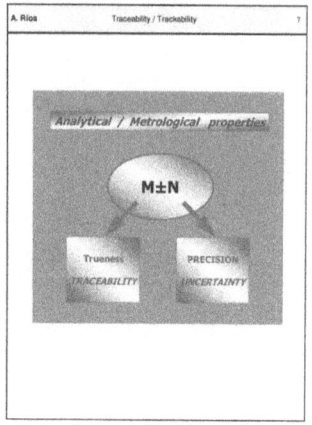
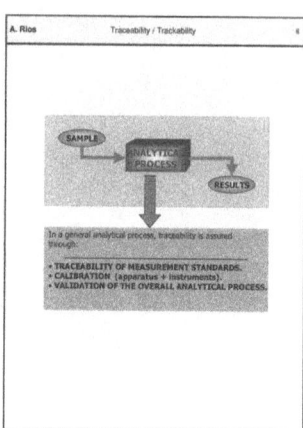
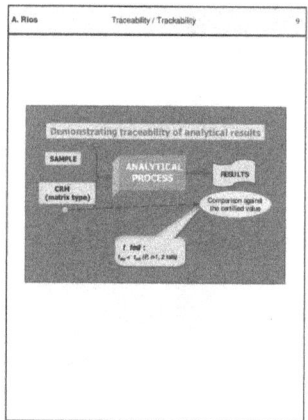
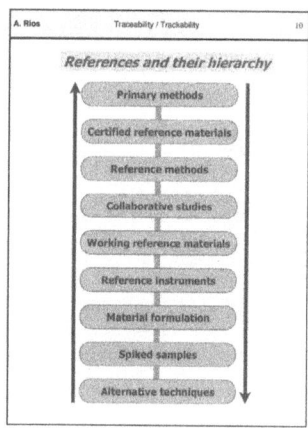
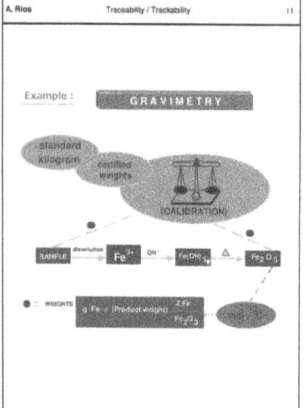
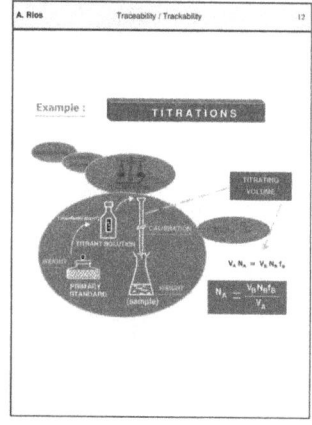

Slides 7-12

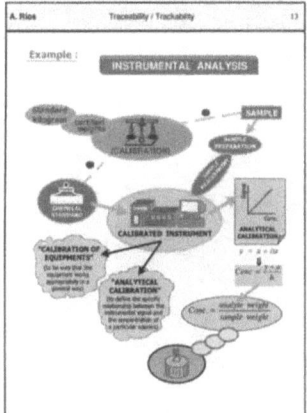
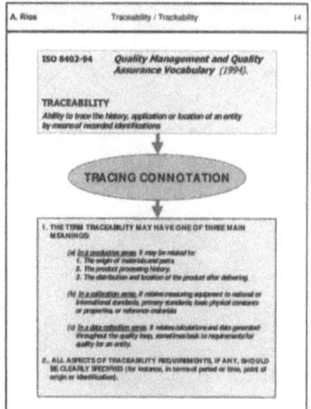
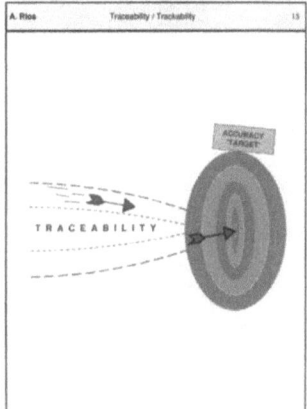
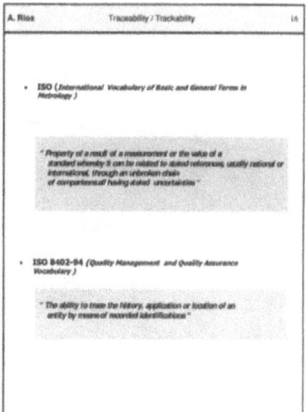
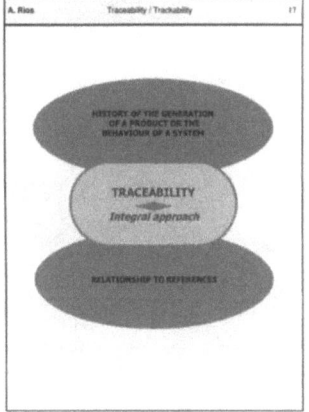
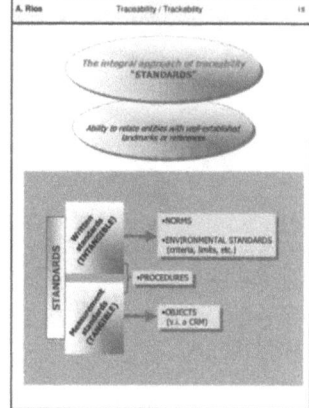

Slides 13-18

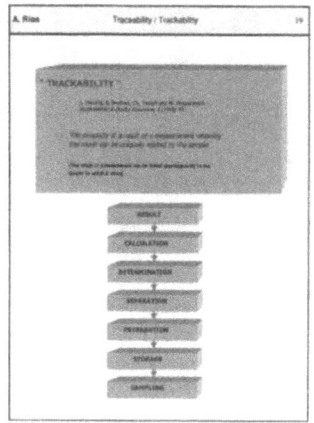
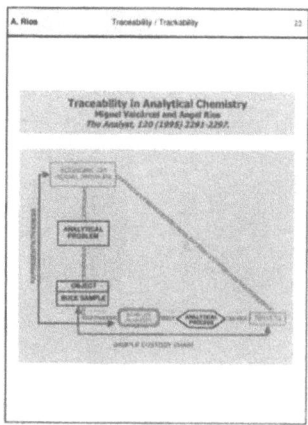

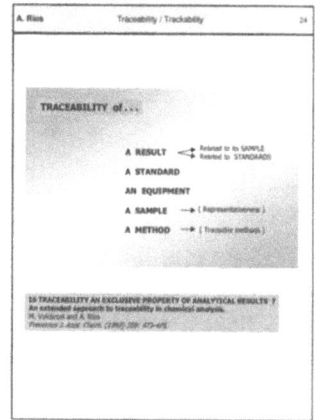
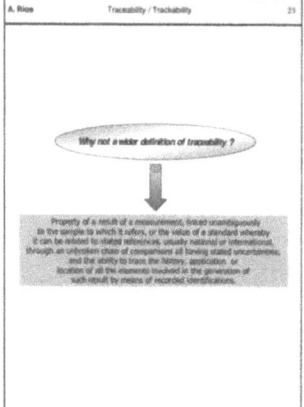

Slides 19-24

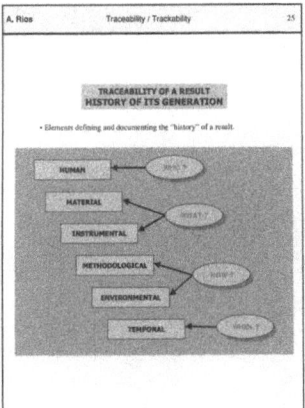
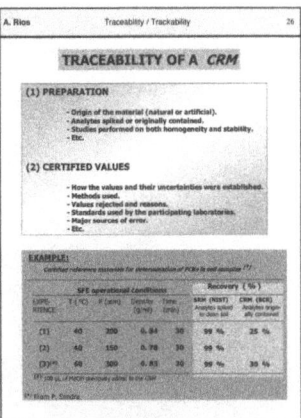
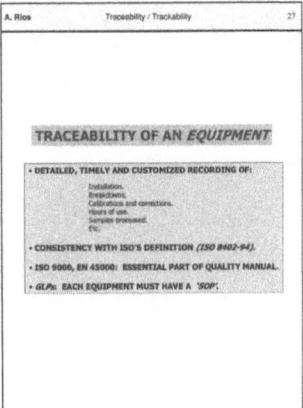
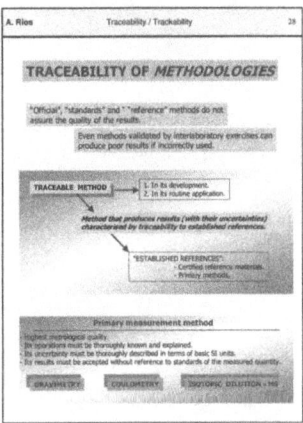
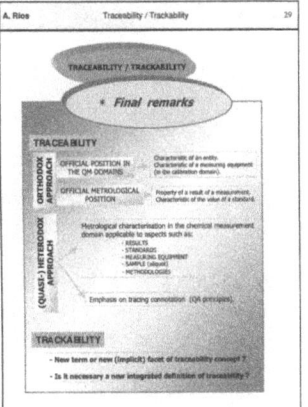
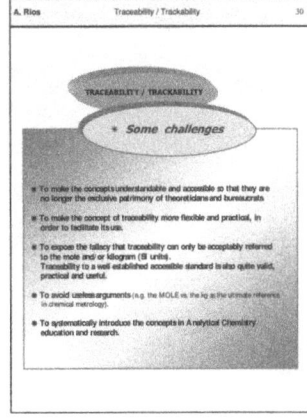

Slides 25-30

Validation: an Example

W. Wegscheider

Abstract
Validation of analytical methods can be regarded one of the most central topics in teaching analytical chemistry. While in the past it might have been feasible to demonstrate principles and practices of the most common analytical techniques to the students, the proliferating wealth of modern methodologies does not lend itself to this traditional approach due to restrictions in time and cost. Irrespective of the experimental possibilities of a particular analytical department, a couple of fundamental abilities and skills will always be presupposed by future employers of young analytical chemists. Among those abilities are critical thinking, problem solving capacity, selection and implementation of analytical strategies.

In implementing validation strategies the stepwise approach from an initial appraisal of a candidate method, to the discussion of alternative methods, selection of minimum quality criteria to meet the requirements of the customer (or scientific goal), the actual laboratory work to provide experimental evidence that these goals are met, to the analysis of the unknown(s) including the final review of the data obtained. If required, this leads to the selection of improved procedures, to the further optimisation of the originally targeted method and to control experiments at the end of the run in order to assure sufficient stability of the measurement system during the entire analytical process.

If organised in this manner, education in validation related matters amounts to a genuinely advanced training in analytical practice that is widely expected from graduates by industry and government. The actual demonstration of quality criteria such as limit of detection, limit of quantitation, working/linear range, selectivity, robustness, sensitivity, repeatability, etc. can add important extra value to the study of the underlying concepts. If validation is taught to undergraduates the examples should be related to relatively simple standards from water (or similar) control, for graduates the interaction between validation and (further) optimisation can be a real challenge.

A worked example is provided by an industrial analytical problem for demonstrating the principles outlined above using the *Excel* macro *"ValiData"* [1] that has widely been accepted by industry as it is based on the pertinent ISO and DIN Standards.

Slide 1
There are several steps that must be mastered in creating a suitably validated analytical procedure that will produce data useful in solving a clients scientific problem. Some understanding of the basics of this problem is therefore indispensible for any analyst.

From this understanding there must come a clear objective for performing the particular measurement prior to validation. From this objective the analyst has to derive the pertinent performance characteristics of the method that will enable her/him to securely meet the objective.

In many cases there will be alternative ways to conduct the measurement, but not all of them will be equally suitable with respect to the level of expertise in the laboratory, with respect to cost, etc. so that one needs to establish priorities, in other words one has to decide which method (or variation thereof) one wants to start with as it promises to lead to the desired results with the least overall effort in time and money.

This then leads to a stepwise plan for executing the experiments necessary to take the final decision regarding the suitability of the tested procedure in the light of the initially established objective.

Slide 2

Good communication and cooperation with the client are at the heart of developing and validating a satisfactory analytical protocol. Once the needs of the client are understood they must be translated into intermediate performance specifications of the procedure that are readily checked during the validation so that no effort is spent in vain. It is not uncommon that there is a critical time frame given within which the data must be available in order to be useful for the client. This time constraint frequently compromises the quality of the data as well as the time that can be afforded for optimization and validation.

A well staffed and well equipped laboratory may also see alternative routes to success for one and the same purpose in which case these must be appraised with an eye on the expected method performance. The reliability of this process will largely depend on the prior experience of the laboratory staff with similar problems, but it is a very necessary step as it makes the analyst aware of the actual level of experience in a particular situation.

For the most promising candidate procedure a detailed validation plan is laid down in writing. It is useful to structure this in a way that will make it possible to judge after each step whether the candidate method is likely to deliver the desired performance in the end, or whether one has to to terminate the validation work, do further optimization or introduce changes into the procedure, or abandon the current route altogether and look for alternatives.

If successfully completed once can proceed with the analytical work on the samples.

Slide 3

The example chosen to demonstrate the validation of an analytical procedure is the control of the maximum allowed content of Fe in very pure magnesium oxide as certain industrial applications of MgO require such a tight control of the Fe content. In the present case the upper limit of iron oxide guaranteed by the supplier of the mineral was 0.5%. The situation was such that the laboratory of the supplier found 0.41% and the laboratory of the buyer found 0.64%.

In an attempt to settle the case laboratory data good enough to distinguish between the two situations are required.

Slide 4
The three relevant values for the content of iron oxide were 0.41, 0.5 and 0.64%, so that there was a difference of roughly 0.1% from one to the next. If one wants to resolve such a difference the overall uncertainty of the analytical procedure including the uncertainty from the sampling process must be significantly smaller that this, say 0.03%. From experience we know that it is generally necessary to allow the larger portion of this overall uncertainty to the sampling process, so it is reasonable to conclude that the uncertainty from the analytical chemical work itself should not exceed 0.01%.

This leads by experience to specific requirements on certain analytical performance characteristics: a limit of quantitation and a limit of detection must definitely be well below the tolerable uncertainty. It was therefore strived for obtaining a limit of quantitation more than an order of magnitude less than the critical analyte level of 0.5% as this at the same time would ensure that the required uncertainty of 0.01% would lead to an acceptable relative uncertainty at the limit of quantitation. If this limit of quantitation were to be no more than 0.03% the relative uncertainty would amount to 33%. As generally the limit of detection is yet lower, this limit should not exceed 0.01% Fe_2O_3 in MgO.

For checking the significance of interferences and thus ensure a sufficient selectivity of the method it was decided to work with certified reference materials as these are readily available in this instance. If not, a more detailed study of possible influences might be necessary. For linearity only a moderate range was required and this was set to cover the range of previous analyses plus a certain safety margin.

Slide 5
In choosing the most promising candidate method one also should consider the expertise in the laboratory, whether it is in routine use or - at least - whether all the necessary equipment is ready, but also the time to complete the validation which must be amended by the time subsequently needed to complete the analysis and also the cost incurring.

Taking all of these considerations into account it was decided to primarily focus on flame atomic absorption spectrometry (AAS).

Slide 6
A validation plan was designed using the standard instrumental conditions for the determination of Fe by flame atomic absorption spectrometry. As the matrix was pure MgO a comparison of calibrations for Fe with and without analytical grade MgO should give a hint towards the presence of interferences from this matrix through a thorough comparison of calibration functions.

These calibration results can at the same time also be used to assess the limit of detection and the limit of quantitation. It is expected that the results in presence of MgO should be fairly representative of those expected in analytical practice.

The residual standard deviation from calibration will permit the estimation of the precision of the procedure after allowing for some variability from sample dissolution and general day-to-day components.

For the recovery study use should be made of the availability of very pure MgO with Fe-content below the limit of detection of the candidate procedure. This material could be mixed in various proportions with an also available certified reference material to produce defined overall contents around the concentration level relevant to this study, 0.5%.

One of the greatest chances to uncover systematic deviations in analytical results is by applying alternative methods to solve the same problem. This is not always simple, but in many instances a method can be found for rough checks. In the present case a titration will be performed.

Finally, it might well be that there are limitations to the validity of interpretation of results that could be caused by the limits to the validity of the sampling process.

Slide 7
The raw data for the assessment of performance characteristics is shown in the table and consists of absorbance data at five distinct concentrations values between 0 and 2mg/L each replicated five times giving a total of 25 readings. The working range defined in this manner spans from 0.5 to 2.0mg/L.

Slide 8
Similarly, five calibration solutions were prepared in the same concentration range but now with the addition of 200mg/L MgO. This amounts to a Fe concentration in the original sample between 0.25 and 1.0%, bracketing the target value of 0.5%.

The readings are very similar and a direct numerical comparison of raw data is futile. One has to resort to graphical and statistical figures in order to produce a meaningful comparison of the two experiments.

Slide 9
Every laboratory has its own expertise and instrumentation at its disposal, but it is good practice to check now all options with respect to their viability. This has to be done by comparing the just defined minimum requirements against all known or expected performance characteristics of all available alternative procedures and spell out the chances for success.

Here, the following methods were considered: gravimetry, titrimetry, graphite furnace atomic absorption spectrometry, flame atomic absorption spectrometry, wavelength dispersive X-ray fluorescence spectrometry and inductively coupled plasma mass spectrometry with a quadrupole analyzer.

Slide 10
The graphical comparison shows a good fit to the regression line in both cases. It is therefore fair to predict that the matrix will not be a severe obstacle to the determination of Fe in the required concentration range.

It is, however, noteworthy that the *absence* of the matrix MgO (left graph) leads to a somewhat curvilinear response that is not observed when measuring in presence of MgO (right graph), so that some differences between the determination of Fe with and without MgO can be expected.

For this purpose it is good to resort to quantitative figures as are provided from regression statistics.

Slide 11

For this purpose an *Excel* macro *"ValiData"* was produced that provides most of the functionality in terms of statistics that is required in method validation. It is programmed according to the provision of the international standard ISO 8466 and the German standard DIN 32645 for regression statistics and limits of detection calculations, respectively.

It starts out with a comparison of standard deviations at the lower and upper end of the working range as these need to be identical for ordinary regression, then it tests for non-linearity according to Mandel´s rules. According to the outcome of these tests there are four different ways to compute the calibration line:

linearity		yes	no
standard deviation	identical	normal regression	quadratic
	different	weighted regression	weighted quadratic

The sensitivity as slope of the calibration line and the confidence as well as prediction limits are also provided.

It is important to note that for method comparison studies more sophisticated variants of regression analysis are necessary. In this macro the two options offered are orthogonal and robust regression.

Slide 12

In the upper left part there are the results in absence of MgO, in the lower right part those in presence of MgO. This statistical analysis confirms the curvilinearity of the calibration line for Fe in absence of MgO.

A direct comparison of sensitivity is not meaningful due to the curvature of the calibration line for Fe in absence of MgO. A very important figure for the present case is, however, the relative standard deviation of the procedure that is pointed at by the two arrows. In both cases it is below 2% in the middle of the calibration range and this makes it likely that the procedure will be useful for the intended purpose.

Slide 13

As there was a requirement formulated with respect to the minimum limit of detection (0.01%) and limit of quantitation (0.03%) these figures can now be extracted from the calibration data according to the cited standards. In absence and presence of the matrix the limit of detection is 0.013mg/L and 0.01mg/L. After taking into account the dilution factor associated with dissolution of the matrix this amounts to 0.007 and 0.005%, respectively. The somewhat lower limit of detection in presence of the matrix is unusual, but could be anticipated in this case due to the better reproducibility and linearity that is apparently induced by the presence of MgO.

The limit of quantitation is only computed for the rectilinear case and amounts to 0.018% well below the previously specified minimum.

Slide 14
For the estimation of trueness a recovery experiment is run with the certified reference material from NIST SRM 104 that contains 7.07±0.04% Fe in MgO. A serial dilution was done with very pure MgO so that the resulting concentration was in the relevant range: six samples between 0.1 and 0.6% Fe in MgO were prepared and measured against the previous calibration curve containing MgO.

The theoretical slope of 1.0 was recovered within statistical limits, therefore there was no indication for a proportional deviation of results from this procedure. However, the intercept was significantly different from 0.0 and gave - 0.009%. Converting this additive deviation to the relevant concentration level of 0.5%, the relative deviation was but -1.8% and thus acceptable for the present purpose.

Slide 15
For an independent assessment a titration was set up for the certified reference material as indicated on the sheet. The deviation was roughly -1% of the certified value and thus also useful for the purpose.

Such an independent procedure is particularly useful when NO certified reference material was available and was applied in the present case mainly for demonstration purposes.

Slide 16
An analytical problem cannot be solved unless the sampling associated with the entire procedure also meets the overall requirements. For Fe in MgO as significant contribution of sampling to the overall uncertainty cannot be ruled out.

Initially, 0.02% Fe in MgO was set as limit to the contribution of sampling to the overall uncertainty. In the present case about 500g have been delivered to the laboratory, most of which is in pieces of roughly 1cm^3. There are pieces of two different colors, the darker ones contain roughly 1% Fe. If one supposes that the light ones contain a negligible concentration of Fe it is possible to compute a worst case estimate of the contribution of sampling to uncertainty.

500g then contain roughly 160 pieces, for a total content of 0.5% only 80 pieces contribute to the Fe-content. This then amounts to a sampling error - if all the material is properly ground, mixed and resampled - of 9%. In terms of Fe-content this is 0.5*0.09 = 0.045% and this clearly exceeds the initially required 0.02%. To achieve a sampling error of 0.02% Fe in MgO (4% rel.) under these circumstances one needs to process 625 pieces of the size of 1cm^3 amounting to 1875g sample in total.

So, although the analytical procedure is adequate for the purpose, one has to specify a minimum sample size of 2kg to meet the expectations of the customer.

References
[1] C. Rohrer, W. Wegscheider, GIT (1994) 38, 688-691

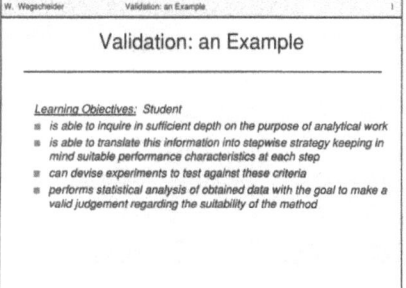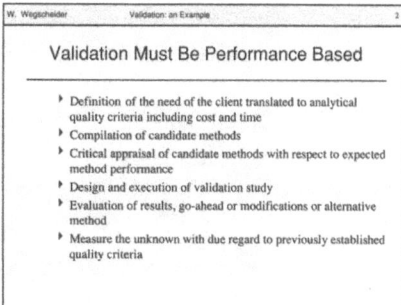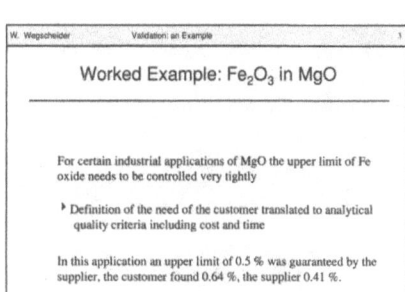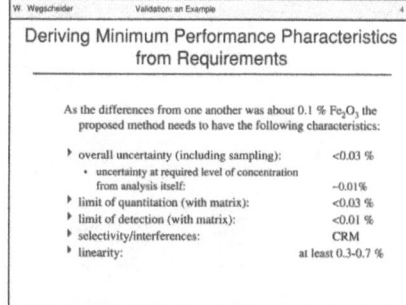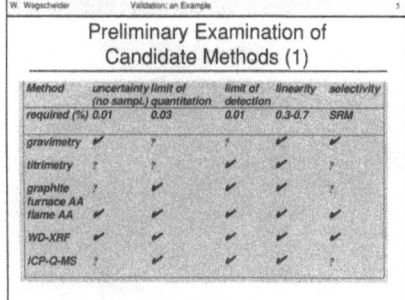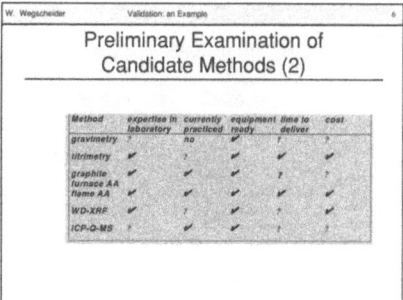

Slides 1-6

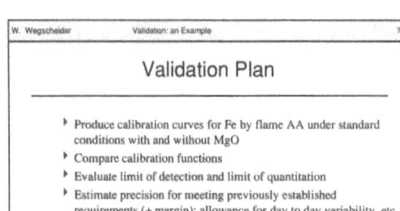
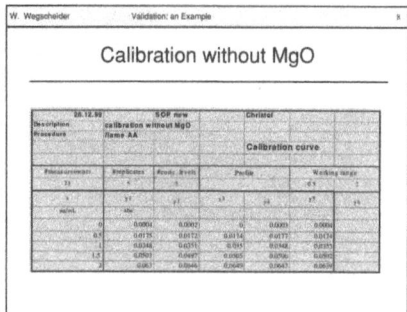
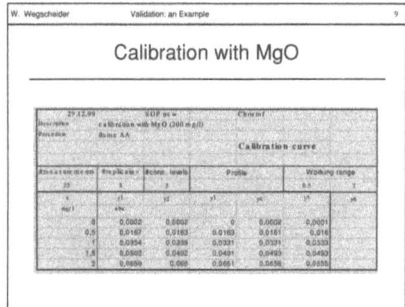
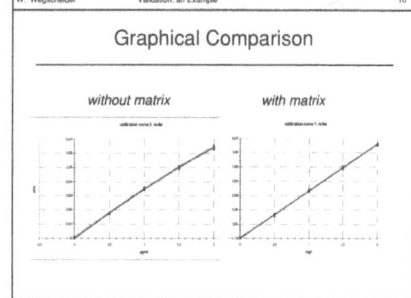
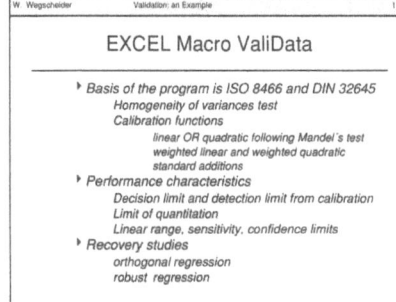
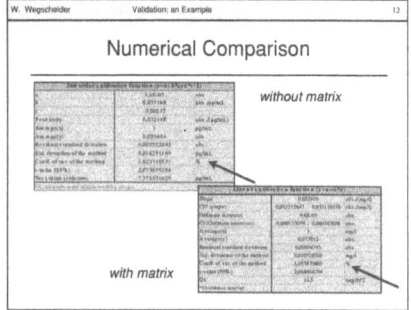

Slides 7-12

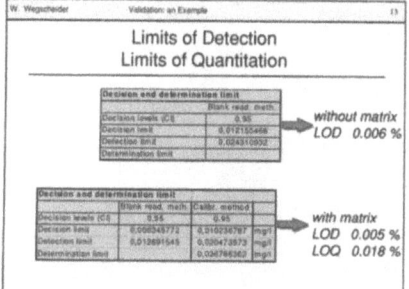
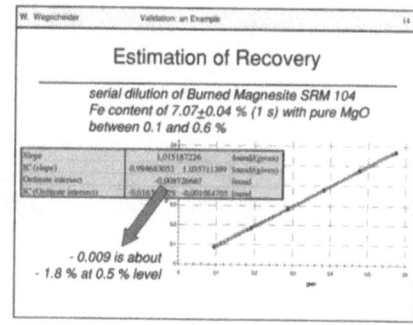
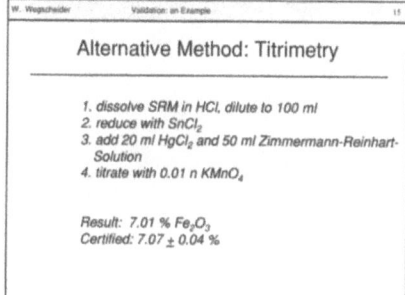
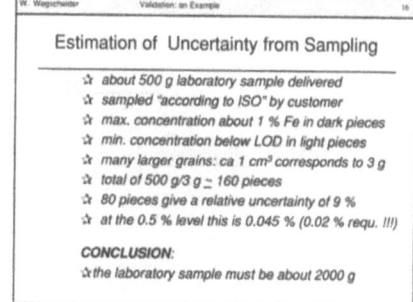
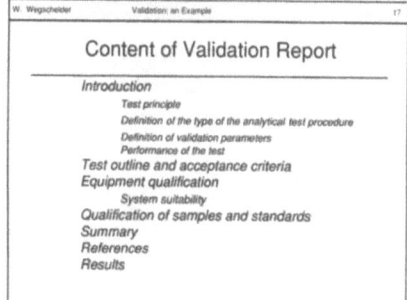

Slides 13-17

Metrology in Chemistry

M. Valcárcel

Abstract
There are two main aspects of the quality of analytical information provided by analytical laboratories. On the one hand, results must be based on metrological principles and, on the other hand, they may be coherent with the requested information by "clients" (solving analytical problems). The first approach will be implemented in a practical way.

Metrology is the science of the measurements. Traditionally, this term has been used to describe measurements of physical parameters, but there is no reason why it should not be used to refer to chemical measurements.

Metrological principles are unique, but the practical connotations of metrology are rather different. There are many important differences between chemical (CMPs) and physical (PMPs) processes which will be emphasised through pertinent examples in order to demonstrate that immediate extrapolations are dangerous and there is a need to carefully adapt the great developments of metrology in physics to the chemical field.

Because most of the written standards and guides are focussed to physical measurements, there is a need to develop documents understandable and useful for analytical chemists which give answers to questions such as "how to be traceable?", "how to validate a method?", etc. In this way metrology will be closer to the chemical bench level.

Slide 1
After placing metrology in chemistry in the Analytical Chemistry quality binomial framework, the general notions of the science of chemical measurements are briefly outlined. A precise description of the most salient features of metrology in chemistry can be achieved by systematically considering the principal differences between physical and chemical measurements. Practical approaches to metrology in chemistry are also emphasized, most of them being the subject matter of other contributions to this text. Some concluding remarks summarizing the pedagogical message end this presentation.

Slide 2
Chemistry rests on four mutually related cornerstones, namely: theory, synthesis, analysis and applications. The best way of representing these basic chemical elements is by using the so-called "chemical tetraedron.

Analytical Chemistry is the science or discipline in charge of one of the chemical cornerstones: Analysis. It is thus the science or discipline of chemical measurements or the chemical metrological discipline. Analytical Chemistry is an information science [1]: it provides more and better chemical information about an object or system by expending less material, time, effort and economic resources.

Slide 3
According to Malissa (private communication), the general aim of Analytical Chemistry is to make the intrinsic but latent information on an object evident, measurable, truthful and useful. This original philosophical approach is based on the use of eight objectives for the word *information* that identify it (chemical), specify its origin (latent, intrinsic), assign it to technical features (evident, measurable, available), and delimits quality in both basic (truthful) and practical (useful) terms.

Slide 4
As an information science, the general objective of Analytical Chemistry is to minimize global uncertainty in the chemical information of an object or system. However, analytical information encompases three different aspects that can be ranked as follows [2]:

At the top of the hierarchy is the inherent, intrinsic, chemical information of the object or system. This information is experimentally inaccessible, represents absolute trueness, is subject to no uncertainty and is consistent with *ideal quality*.

One step down is the information regarded as true (for example, certified values for a CRM), which is experimentally accesible only in exceptional circunstances (e.g. values from interlaboratory excercises). This information corresponds to *referential quality*, the highest level of accuracy and the lowest specific uncertainty.

On the third hierarchical level is the information (results, reports) routinely produced by laboratories, which corresponds to *real quality* and exhibits low accuracy and maximum uncertainty.

Slide 5
Quality can be approached in two complementary ways. From a basic point of view, it materializes in a set of attributes, characteristics or features. On the practical side, quality entails meeting the "clients" requirements.

Analytical Chemistry also possesses the corresponding basic and applied facets as regards quality of the analytical information, namely: analytical properties [3] and solving the so-called "analytical problem" [4], respectively.

There is thus full correspondence between the two sides of the Analytical Chemistry quality binomial. Metrology in chemistry can be assigned to the basic facet, whereas consistency between requested and delivered chemical information is involved in the applied aspect of the binomial.

Slide 6
Assuring quality always involves comparing and comparing entails using a reference. Properly define the two facets of the Analytical Chemistry quality binomial, which can be described as analytical quality, one must establish pertinent references. Measurement standards and written standards (norms, guides, protocols, etc) are the references for metrology in chemistry. Properly solving analytical problems entails knowing the features of the information needed by the clients and using written standards as references.

Slide 7
Both sides of analytical quality are essential ingredients of the quality of the analytical information provided by laboratories. Integral analytical quality can be represented by using an hydrodynamic model. Analytical Quality is fed by two streams of metrological quality and consistency between delivered and requested analytical information. Metrological quality, in turn, contains two essential, mutually related ingredients, viz accuracy and low uncertainty, which are fed by the philosophy behind traceability [5, 6]. The consistency between delivered and requested information is fed by a capital property (representativeness) and the principles of "fitness for purpose", both under the philosophy of the analytical problem.

Several other components of the quality of analytical information are related to both sides (e.g. proper sampling and complementary analytical properties such as expeditiousness and reduced costs).

Slide 8
In summary, metrology in chemistry is the responsibility of Analytical Chemistry and it is strongly related to its foundations and to the quality of analytical information. However, it is not the sole goal of Analytical Chemistry, but rather is indirectly related to solving the analytical problem.

Slide 9
Metrology (from "metron", the Greek for "measure") is the science of measurements. While this word has traditionally been used in connection with measurements of physical parameters such as time, temperature, pressure, dimensions, etc, there is no reason why it should not be used with measurements of physico-chemical (e.g. reaction rates, equilibrium constants, ionic strengths), chemical (e.g. presence/concentration of chemical species in complex samples), biochemical (e.g. enzyme activities, aminoacids from proteins) and biological parameters (e.g. microbial or aerobiological counts).

Analytical Chemistry is mainly concerned with chemical, biochemical and physico-chemical measurements. However, in special cases, some physical (e.g. weights, temperature) and biological measurements (e.g. microbial counts) are also the responsibility of analytical chemists. Such is the case with the characterization of the quality of waters.

Slide 10
Metrology in chemistry is thus the science of (bio)chemical measurements. It shares common basic principles with metrology in physics and metrology in biology, but the practical connotations of these branches of metrology are quite different.

Slide 11
Physical, chemical and biological measurements have the same metrological foundations but their applied facets are rather different.

Slide 12
These differential practical connotations arising from the nature of the measured parameters are reflected in two essential metrological aspects.

On the one hand, the difficulty of implementing reliable measurements grows from physical to biological parameters. On the other, availability of proper measurement references decreases dramatically from metrology in physics to metrology in biology, with metrology in chemistry in between chemical measurements are more difficult to implement than physical measurements but easier than biological measurements.

Slide 13
The significance and impact of metrological principles in Analytical Chemistry is undeniable. It is possible to distinguish between generic, basic and specific trends in Analytical Chemistry according to significance and scope. One of the three generic trends is the exploration of its metrological character, a basic aspect of Analytical Chemistry that was not systematically considered until two decades age. Among basic trends, the development and proper use of reliable measurement references is crucial.

Slide 14
Switching from a classical to a metrological approach to Analytical Chemistry entails philosophical, strategic and specific changes that involve a positive evolution to basic analytical quality. Although no substantial innovations are introduced, Analytical Chemistry is both scientifically and technically enriched as a result. In fact, the old principles do not dissapear; rather, they are improved.

Slide 15
Expanding the metrological character of Analytical Chemistry implies completing and replacing accuracy and precision (mutually related) with traceability and uncertainty, respectively, which are also strongly related to each other.

Standard, official, reference methods in traditional Analytical Chemistry are replaced by traceable and primary methods in the metrological approach.

Slide 16
Considering the metrological character of Analytical Chemistry also involves three major changes, namely:
- from global, non-specific interest on analytical processes to greater emphasis on both measurement references and quality of measurements
- from a non-systematic, generic interest on quality to the implementation of quality systems in the laboratory, i.e. the development of Quality Assurance, Quality Control and Quality Assessment activities and
- from paying little attention to other laboratories to establishing fruitful interlaboratory excercises in order to achieve both comparability and harmonization with a view to the mutual recognition of results, a key aspect of social and economic relationships.

Slide 17
Establishing the most salient differences between chemical measurement processes (CMPs) and physical measurements processes (PMPs) is an excellent way of outlining the intrinsic features of metrology in chemistry [7].

Furthermore, these differences allow analytical chemists to learn from and to take advantages of the impressive developments of (traditional) metrology in physics in order to carefully adapt them to metrology in chemistry.

This systematic comparison can also be used to demonstrate the need for specific development of metrology in chemistry. Immediate extrapolations are generally wrong and dangerous. One must to take into account that norms, guides and dictionaries of metrology are strongly reliant on physical measurements as a consequence of the longer tradition of physical metrologists. This orientation is a serious hindrance to the systematic expansion of the metrological character of Analytical Chemistry.

Slide 18
CMPs are strongly sample-dependent, by contrast, PMPs are virtually sample-independent. For example, similar length transfer standards can be used to measure the length of a table, the height of a tree or the distance between two objects. On the other hand, CMPs and their measurement standards are quite different when applied to different types of sample, even if the measurand is the same. Such is the case with the determination of a pesticide in different types of water (from spring to waste water), soil, sediments, commercially available formulations, biological fluids, vegetables, etc.

Slide 19
The combination of millions of different chemical species and thousands of sample types can result in an enormous variety of analytical problems. CMPs are thus extremely varied, which poses a severe practical limitation and restricts the number of general procedures available. By contrast, PMPs are more general and more widely applicable.

Slide 20
The need to implement preliminary operations such as sampling, sample treatment, etc, involved in most CMPs is of great practical significance because they strongly influence the quality of the results and are tedious, time-consuming and hazardous is some instances. By contrast, preliminary operations are of little significance in PMPs. As a result, traceability in PMPs is more immediate and easier to establish.

Slide 21
Measuring equipment is not the sole focus in CMPs (PMPs rely almost exclusively on direct-use instruments). In CMPs, the quality of the results is affected by many other factors.

Slide 22
Sampling is a key factor with a view to achieving representativeness in chemical and biological results, In general, this operation does not exist or is quite simple in the physical field. Moreover, the heterogeneity of collected samples introduces serious problems at the sub-sampling stage. The problem is almost exclusive of metrology in chemistry and biology.

Slide 23
The need for appropriate measurement standards in metrology in chemistry has been meet by only 5-10% (~1% in biology) as opposed to 80-90% in physical measurements.

Taking into account that a) Analytical Chemistry without quality references makes no sense, b) both calibrants (as in PMPs) and matrix-type (certified) reference materials are needed, and c) the impressive number of references needed to encompass the millions of analytes and thousands of sample type possible; the development of standards in metrology in chemistry is a hot research topic in Analytical Chemistry. Because of these limitations, there is a need to develop reliable alternatives to establishing appropriate measuring standards in order to assure the metrological quality of the results; these technical approaches range from adapted calibration procedures (e.g. the standard addition method) to the participation in inter-laboratory excercises, a detailed description of which is beyond the scope of this presentation.

Slide 24
Unlike PMPs, measuring standards in CMPs are not exclusively used for calibrating instruments, but also for a variety of purposes such as global evaluation. For instance, calibration in CMPs comprises two mutually dependent, namely: operations equipment calibration (similar to PMPs) and analytical calibration (which does not exist in PMPs).

Slide 25
Traceability is rather different in metrology in chemistry than in metrology in physics. PMPs are more easy to trace because of the wider availability of transfer standards and the simplicity of the processes involved. By contrast, traceability in CMPs is a metrological characteristic that is difficult to demonstrate if the orthodox approach described in the metrological dictionary [8] is strictly adhered. The ordinary last reference in the unbroken chain of comparisons supporting the traceability of a result is an SI unit in metrology in physics. In metrology in chemistry, the kg and the mole can also be the last reference in the traceability chain; however, this approach is rather distant from the bench top level. Other references can be used as alternatives to the SI units.

Slide 26
The standards used in metrology in physics are in general quite homogeneous; such is the case, for example, with transfer weights (from the laboratory standard to the kilogram). On the other hand, the standards used in metrology in chemistry are quite rather heterogeneous.

There are three general groups of standards relevant to metrology in chemistry that differ markedly from one another. They are hierarchically related according to two contradictory criteria, namely, nearness to the value held as true and availability. Basic standards such as the kg or the mole (SI units) are at the top of the hierarchy and constitute the ultimate references for both physical and chemical measurements. Chemical standards (e.g. atomic weights, the Faraday constant, isotope 12 of carbon, Avogadro's number, ultrapure (99.999%) silver) can be regarded as the traceability links between SI units and analytical standards used in practice, which, in turn, can be classified into two groups (primary and secondary).

Slide 27
The heterogeneity of the standards used to establish the traceability links in

metrology in chemistry can be illustrated with a simple titrimetric example: the determination of chloride in waters. Sodium thiosulphate is the titrant; however, it is a secondary (working) standard so factoring (the link between primary and secondary standards) is needed. Factoring is done in two steps the result of which is a titrant solution of $S_2O_3^{2-}$ of accurately known concentration. The analyte must first react with excess I^- to stoichiometrically produce I_2, which is titrated by the $S_2O_3^{2-}$ solution. The traceability chain thus includes several types of chemicals (tangible standards) and atomic weights (intangible standards).

Slide 28
The availability of transfer standards supported by a well-established national and international network of physical laboratories, all directly or indirectly related to BIMP, is one great advantage of metrology in physics that in fact does not exist in metrology in chemistry.

Slide 29
According to Horwitz [9], uncertainty in CMPs and PMPs is approached rather differently in practice. Thus, physical measurements and chemical measurements have entirely different error patterns that behave differently on replication. Correctable local bias predominates and random errors are smaller in physical systems. By contrast, random errors predominate and bias is difficult to identify and erradicate in chemical systems. Monitoring CMPs entails randomizing in the interlaboratory environment, a concept not considered in the conventional ISO treatment of uncertainty biased by physical metrologists. It is thus necessary to develop new alternatives for estimating uncertainties in metrology in chemistry.

Slide 30
Qualitative analysis is a distinct aspect of metrology in chemistry because it is virtually absent from metrology in physics.

The binary yes/no response is gaining significance in Analytical Chemistry owing to the pressing information demands posed by clients and to the development of powerful qualitative analytical tools such as responsive analytical systems (sensors, screening systems), efficient column separation techniques (GC, LC, capillary electrophoresis), identification techniques (MS, FT-IR, NMR) and hyphenated techniques (GC-MS, LC-MS).

The absence of references in metrological norms and guides to qualitative analysis opens up an interesting prospect: analytical chemists must develop specific documents supporting the metrological side of the binary response.

Slide 31
The most salient practical approaches to metrology in chemistry can be outlined by considering a) the framework of the Analytical Chemistry quality binomial, b) the intrinsic features of metrology in chemistry, and c) the differences between CMPs and PMPs.

Slide 32
The specific practical approaches to assuring quality in chemical measurements include sampling, uncertainty calculations, calibration, last reference in the traceability chain, validation, internal quality control and proficiency testing.

Some are only briefly commented on because they are the subject matters of other contributions to this book.

Slide 33
As stated before, sampling is a preliminary operation in CMPs that can significantly contribute to the uncertainties of analytical results. Unfortunately, it is virtually ignored in metrological norms and guides. There is a need to systematically consider such a difficult operation as substantial part of the quality schemes applied in analytical laboratories.

Slide 34
Traceability cannot be properly defined without considering uncertainties according to the definition contained in the ISO metrological dictionary [8]. The presence of so many sources of uncertainty in CMPs relative to PMPs makes calculation of uncertainties a severe problem that requires immediate addressing. The traditional metrological approach [10] (so called "bottom-up" or "step-by-step") is based on the identification of all sources contributing to dispersion and on calculating and combining the quantified individual uncertainty components. Although this approach has been adapted to chemical analysis [11], it is tedious, time-consuming and analytically unrealistic (*viz.* distant from the bench top level). There is thus a need to develop specific procedures for estimating uncertainties. Several such procedures have been proposed in recent years; all aim to simplify the conventional approach by combining uncertainty sources.

Slide 35
The process for calculating uncertainties using a global approach [12] is depicted in this slide. The total variance (U^2_{total}) is the summation of the variances resulting from two main sources, namely:

a) the analytical process as a whole. The corresponding uncertainty ($U_{process}$) is calculated by chemometric treatment of the results provided by several independent analyses of aliquots of the same CRM, taking into account the uncertainty associated to the certified values; and

b) sample diversity and heterogeneity ($U_{samples}$), the contribution of which is calculated by processing replicates of different samples (S_1, S_2, ..., S_n).

In this way, all uncertainty sources are combined and considered in a global manner. The procedure is quite simple relative to the traditional (physical) metrological approach.

Slide 36
Calibration is a key cornerstone for metrology. In fact, it is an essential aspect of analytical quality and a relevant element of the unbroken chain of comparisons implicit in the traceability concept. It has direct influence on the uncertainty associated to an analytical result. Also it is a substantial ingredient of method validation. All these characteristics of calibration are shared by metrology in physics and chemistry. In practice, however, this metrological operation is quite different in PMPs and CMPs. The most immediate difference arises from the availability of measurement standards. The special features of CMPs allow to distinguish between two types of calibration in metrology in chemistry and to

establish a classification of CMPs.

Slide 37
Metrology in chemistry includes two different but complementary notions of calibration that exist only in a few cases in metrology in physics. Equipment calibration entails ensuring that an instrument or apparatus operates correctly, i.e. as expected. It is intended to correct an instrument's response or parameter indication of the functioning of an apparatus until the assumed true value, which is assigned to the value of the standard employed, is reached. The standards currently used for instrumental calibration are reference materials containing no analyte. Instrumental calibration involves analytical information (that directly related to the results) and non-analytical information (that related to the functioning of apparatus and devices).

On the other hand, methodological analytical calibration involves characterizing the response of an instrument (never an apparatus) as a function of the properties of one or more analytes. It establishes an unequivocal, reliable relationship between the instrument signals and the presence/concentration of the analytes. In quantitative analysis, a calibration curve is obtained as result. The analyte is contained in the analytical standards used for such a purpose, except in titrimetry, where the factor of the titrant solution is established from an analytical standard which does not contain the analyte but has similar characteristics. The information produced is quite relevant to the calculation of the final results (the analytical results).

Instrumental calibration is used both in PMP and CMP, while methodological calibration is used mainly in CMP.

Slide 38
The differences between instrumental and methodological calibration can be illustrated by using a simple example from UV-visible spectroscopy. The target of equipment calibration is the spectrophotometer itself, which can be checked both quantitatively by using absorbance standards (e.g. a potassium dichromate solution that can be an RM or CRM), and qualitatively, by using wavelength standards such as the holmium filter. The target of methodological (analytical) calibration is the chemical measurement process that uses the spectrophotometer as instrument. The standards used are solutions of increasing concentration of the analyte that provide the well-known calibration curve. This type of calibration is rather unusual in physical measurements.

Slide 39
The different types of analytical calibration and the standards involved allow to establish a classification of quantitation methodologies in metrology in chemistry that is shown here. It is based on the ISO approach but includes some changes.

Slide 40
As stated before, from an orthodox point of view the last reference in the traceability chain is the same (SI unit) in CMPs and PMPs. However, the difficulties inherent in accomplishing traceability in metrology in chemistry have raised the need for alternatives to the orthodox approach. The results of CMPs can

be traced to well-established, well-stated measurement standards (CRMs) or to a primary method, a centre of excellence (reference laboratory) or a group of laboratories participating regularly in interlaboratory excercises (proficiency testing).

Slide 41
Validation plays a pivotal role in accomplishing and ensuring quality of analytical results. Both internal and external quality control activities should be systematically considered to this end. Validation also involves the two facets of analytical quality: basic (metrology quality) and applied (fitness for purpose).

Although "method calibration" is the most prominent group of activities in this context, other analyical tools can also be subjected to a validation process. In order to avoid confusion, one must establish a hierarchy of analytical validation such as that depicted here.

There are several categories of validated methods: in house, peer reviewed, collaborative studies to international guidelines, and programmes for test kits. The validation category required by a laboratory is dictated by the type of analysis, the reason for performing the analysis and the intended use of the analytical data. Very many documents and publications exist on method validation. In general, validating methods involved considering the following: scope of the method, accuracy (trueness + precision according to ISO 5725), sensitivity, selectivity, linear range, limit of detection, limit of determination, recovery and ruggedness.

Slide 42
The orthodox approach to metrology in chemistry is coherent with chemical principles in general and with the Analytical Chemistry quality binomial, in particular. Also, the basic principles of metrology in chemistry must coincide with those of metrology shared by Physical metrology. On the other hand, metrology in chemistry must be heterodox as regards the practical connotations of (traditional) metrology in physics.

Slide 43
Consolidating metrology in chemistry entails undertaking political actions and research activities in the following direction:
- There is a need to develop specific norms, guides, protocols and dictionaries adapted to the intrinsic practical connotations of metrology in chemistry. Direct application of existing written metrological standards is extraneous to routine work in chemical laboratories.
- There is an urgent need to increase the availability of measuring standards for chemical measurements in the form of both calibrants (inorganic, organic, organometallic) and matrix-type (CRM) standards for the wide variety of CMPs possible.
- It is also necessary to find reliable alternatives to unavailable measurement standards in order to assure metrological quality in analytical results.
- Considerable efforts should be made to establish traceability for the most common CMPs by considering realistic alternatives for the last standard of the unbroken chain of comparisons. It is also important to raise awareness of the importance of traceability among analysts and clients.

- More rational procedures for estimating uncertainties in chemical measurements based on chemometric approaches are urgently needed.
- Sampling must be regarded as a decisive activity in metrology in chemistry.
- It is also necessary to devote R&D efforts to developing new methods of measurement, new instrumentation and new sample treatment procedures in order to improve quality of the results provided by analytical laboratories.
- It is very important to promote validation activities, which are crucial in metrology in chemistry. Method validation must be approached in three complementary ways, namely: a) in-house; b) through participation in proficiency testing schemes; and c) by developing validated methods through collaborative studies.
- Metrology in chemistry must be approached to the binary response of qualitative analysis, a topic of current and future interest. This area necessitates strong basic support to assure the quality of the information provided by the specific analytical tools designed for each purpose.
- Existing centres of reference working in specific areas of chemical analysis should be recognized and established new ones. A network linking such laboratories and routine-type laboratories could be the most effective way of consolidating metrology in chemistry.

REFERENCES

[1] M. Valcárcel, *Trends Anal. Chem.* (1997) 16, 124
[2] M. Valcárcel and A. Ríos, *Trends Anal. Chem.*, (1999) 18, 68-75
[3] M. Valcárcel and A. Ríos, *Anal. Chem.* (1996) 65, 781A
[4] M. Valcárcel and A. Ríos, *Trends Anal. Chem.* (1997) 16, 385
[5] M. Valcárcel and A. Ríos, *Analyst* (1995) 120, 2291
[6] M. Valcárcel and A. Ríos, *Fresenius J. Anal. Chem.* (1997) 359, 673.
[7] M. Valcárcel, A. Ríos and E. Maier, Accred Qual Assur (1999) 4, 143-152
  Summary of the final report of SMT4-CT96-6505 EU project on
  Metrology in Chemistry and Biology)
[8] International Vocabulary of Basic and General Terms in Metrology,
  2nd edition (1993) ISO,. Geneva
[9] W. Horwitz and R. Albert, *Analyst* (1997) 122, 605
[10] Guide to the Expression of Uncertainty in Measurements, ISO, Geneva (1995)
[11] Quantifiying Uncertainty in Analytical Measurements, version 6, EURACHEM (1995)
[12] A. Ríos and M. Valcárcel, *Accred. Qual. Assur.*(1998) 3, 14

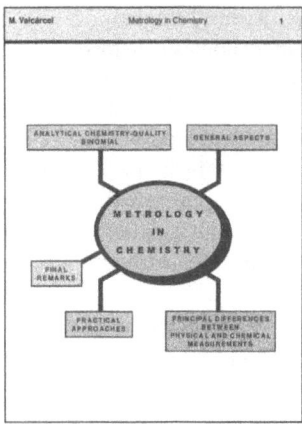
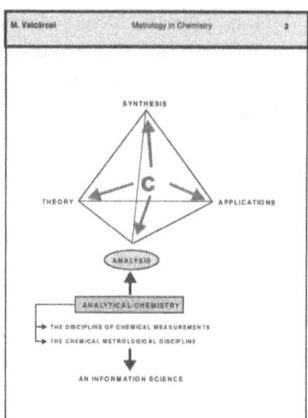
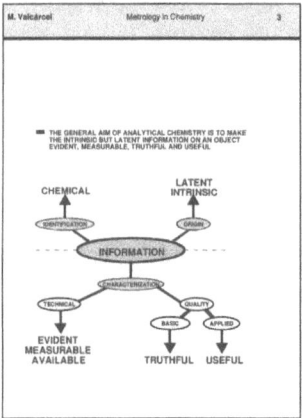
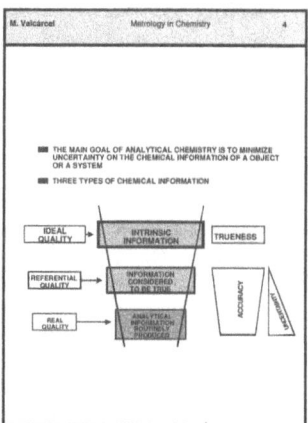
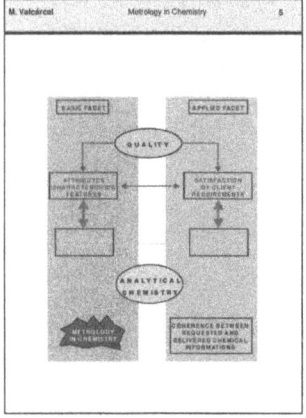
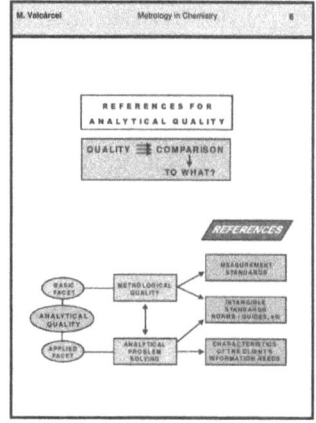

Slides 1-6

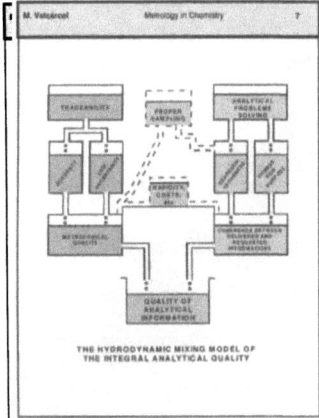
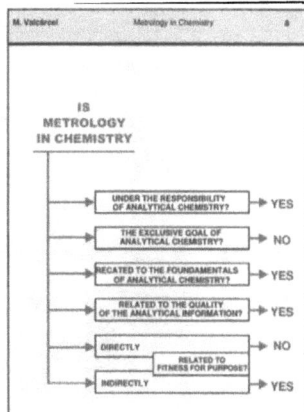
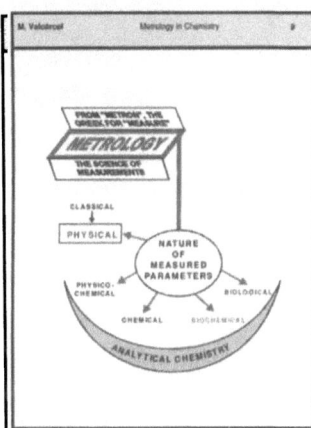

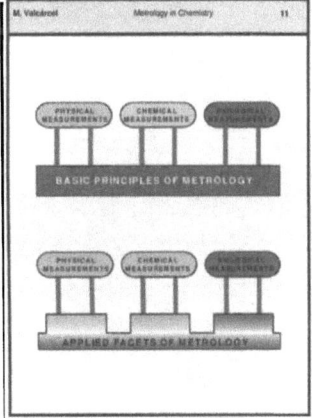
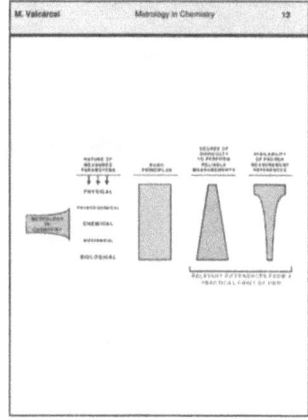

Slides 7-12

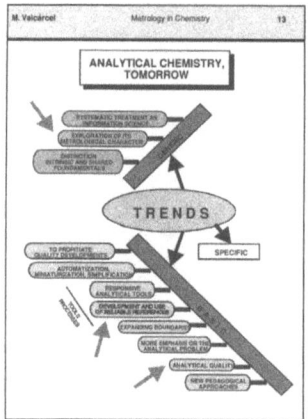
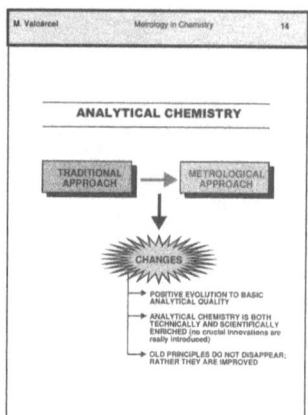
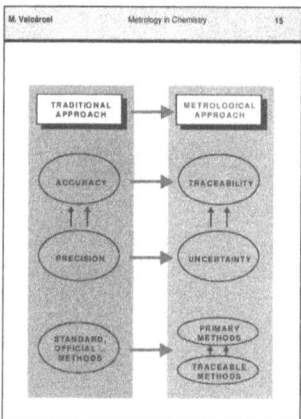
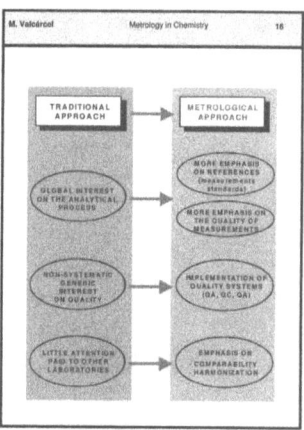
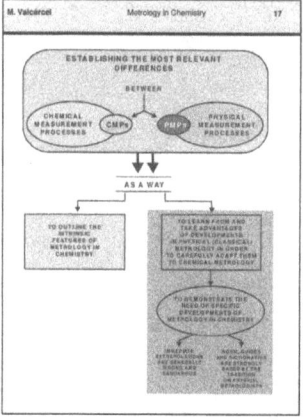
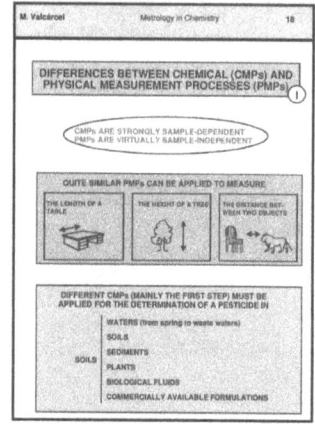

Slides 13-18

Slides 103

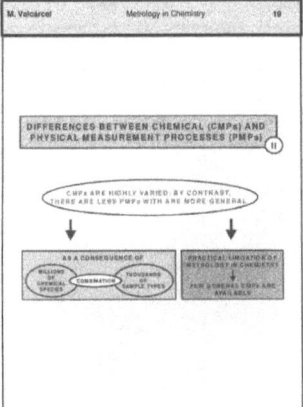
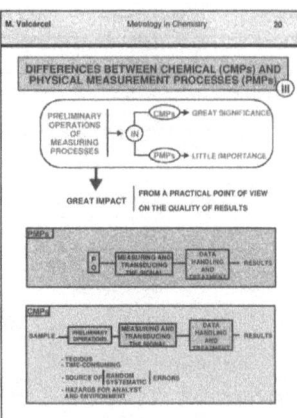
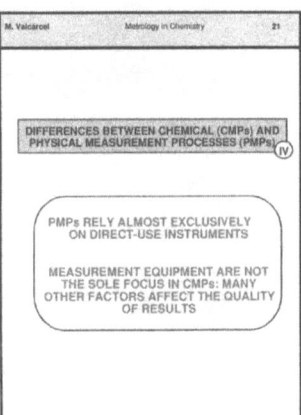
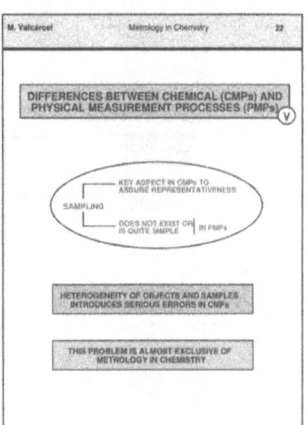
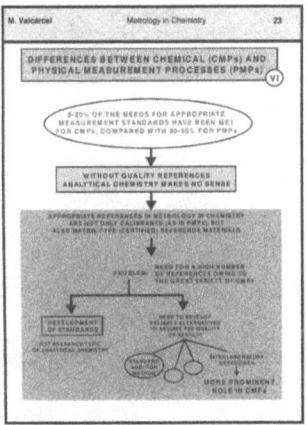
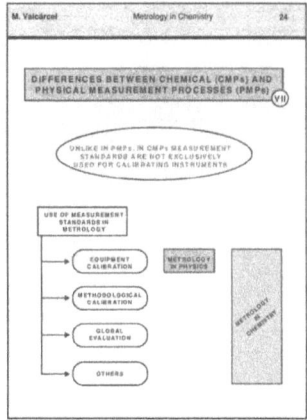

Slides 19-24

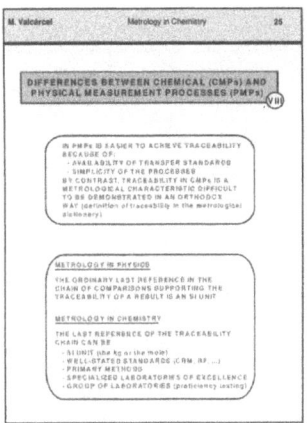
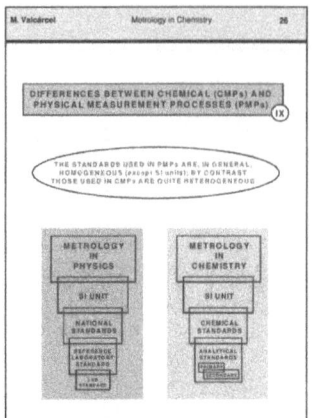
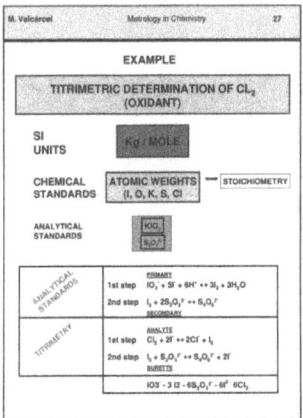
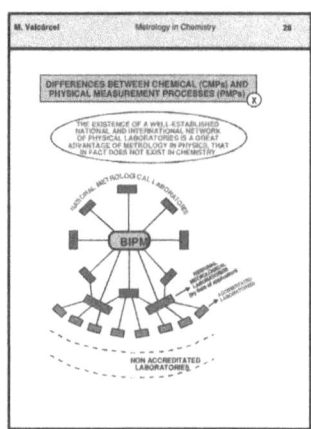
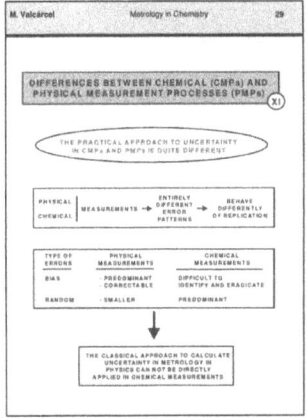
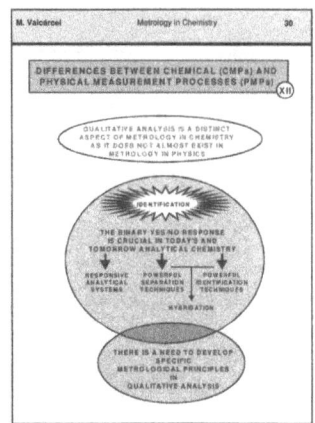

Slides 25-30

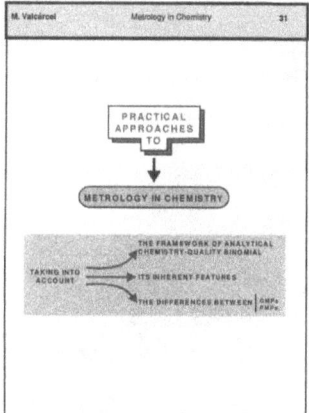
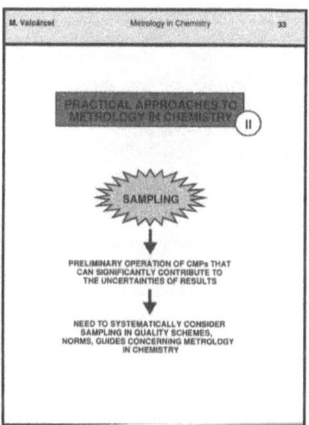
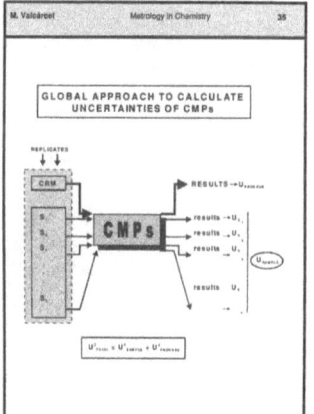
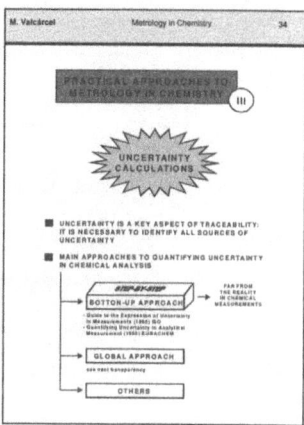
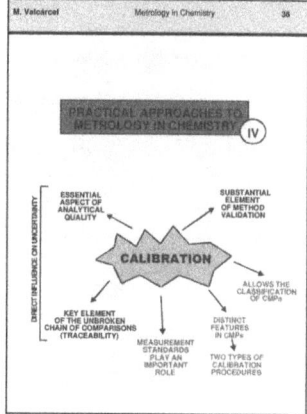

Slides 31-36

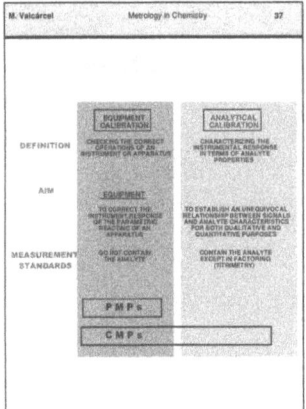
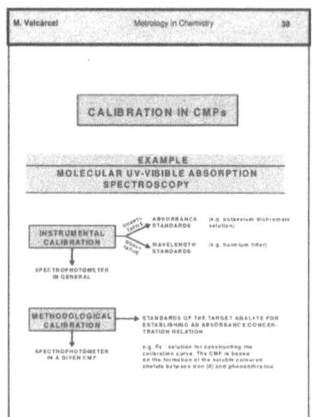
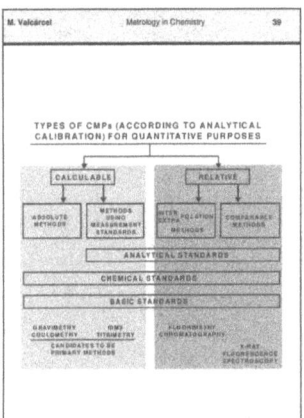
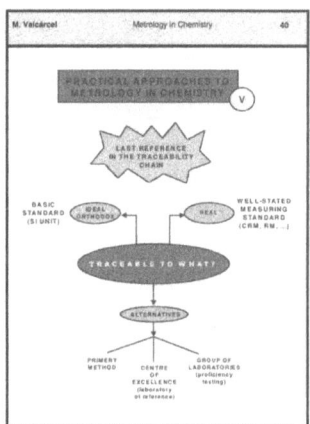
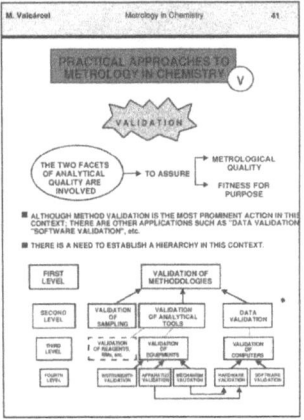
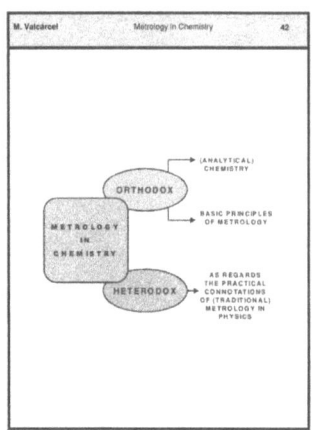

Slides 37-42

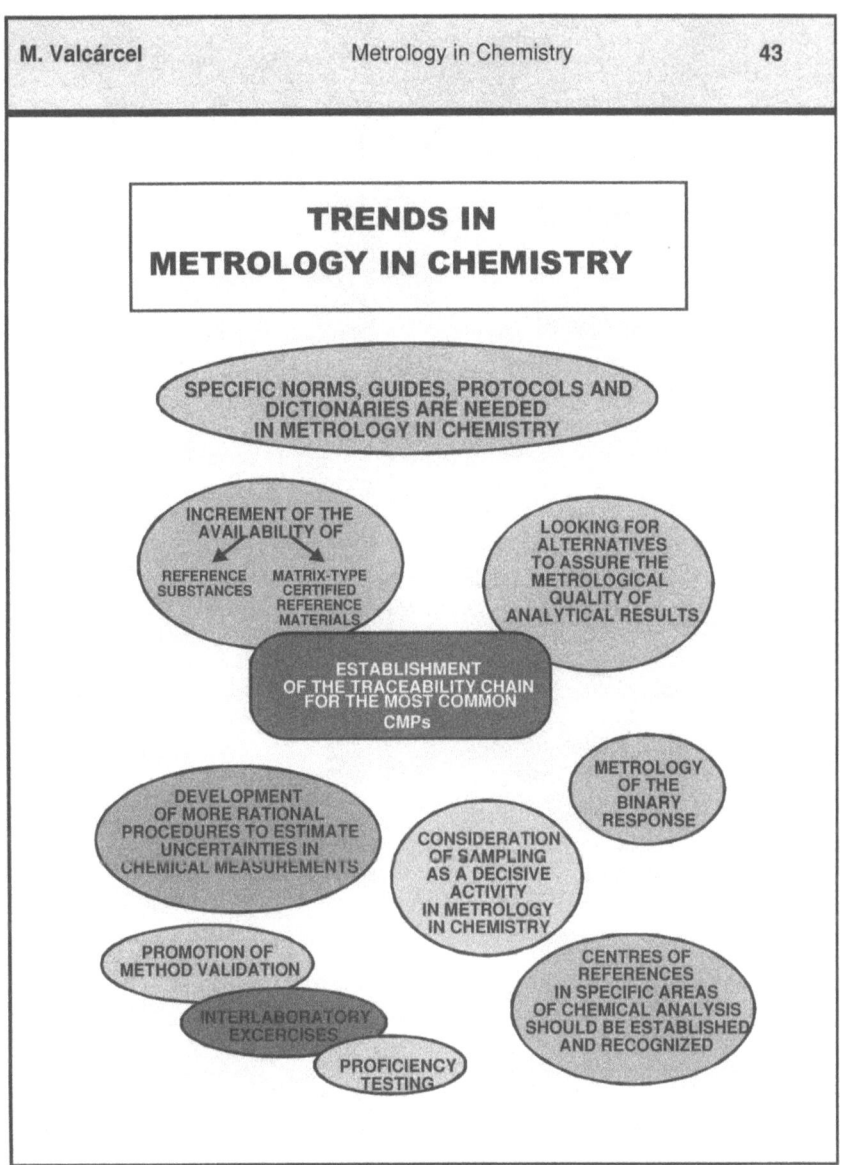

Experiments as Tools to Demonstrate Principles of Quality Assurance

Basic Course Experiments to Demonstrate Validation

H. Albus

Abstract

Undoubtedly, nowadays modern instrumental Analytical Chemistry (AC) is one of the most important interdisciplinery sciences. Therefore teaching AC at university should have a very high priority - also in undergraduate and graduate courses. In this connection both fundamental analytical strategies - accurate results and obtaining their quality - should be implemented from the very beginning into the chemistry curriculum. One approach in this context is the introduction of fundamental analytical terms in such a manner that the results of measurements performed by students are used to highlight a specific analytical problem and the concept - condensed in the selected term - to solve it. One pivotal analytical problem can be formulated as "How can one be sure that a method produces accurate, i.e. precise and true results?". This problem is embedded in the context of the term "validation of a method" which is linked with the question „What has to be done to assure that a method is fit for the intended analytical purpose?"

The ability of an analytical method to produce true results can be evaluated by means of the application of independent comparison methods. In this case subsamples of a sufficient homogeneous material are analysed using the method under investigation and methods based on different measurement principles. To illustrate this approach a simple experimental arrangement was worked out utilizing gravimetric determinations of inorganic ions (Fe^{3+}, Al^{3+}, SO_4^{2-} and PO_4^{3-}) as methods under investigation and photometry resp. flow injection analysis as independent comparison methods. In order to avoid the use of a „black box" method and to enhance the learning effect a "didactic" photometer was constructed which enables the students to have a close look at all relevant components of the instrument and to realize the measurement process in detail. The whole exercise is aimed at students in the first year and can be performed e.g. within the first laboratory course. The students can work together in small groups and in a subsequent tutorial the obtained data can be used to draw conclusions about the specific performance characteristics of the employed methods.

Slide 1

Undoubtedly, modern instrumental Analytical Chemistry (AC) is one of the most important interdisciplinery sciences. Therefore teaching AC at university should start as soon as possible - consequently in the first practical (basic) courses the students have to attend. This presentation shows how simple it can be to explain and demonstrate fundamental analytical terms and techniques by applying appropriate experimental arrangements. Additionally, these arrangements allow the students to actively work *and* experiment with the instrumental equipment in order to get an in depth view of how modern analytical chemistry works in practical terms.

Slide 2
Obviously, it is not possible to explain the basic concepts of AC in complete within a single lesson unit embedded in a basic course in analytical chemistry. The task of a unit like the one presented here should be to make basic terms of AC accessible to the students and to show them how useful and indispensable these are. In this connection two of the fundamental analytical strategies - measuring accurate results and describing their quality - combined in the concept of method validation, are demonstrated using simple, robust and efficient modern analytical methods.

Slide 3
The term "good result" is widely used in the scientific world and at first sight the meaning of this word seems clear. A good result is a result which satisfies... what? Or whom? By asking these questions it becomes clear that "good" is subject to a definition adapted to the given problem! There is no "good" result unless one has defined in detail which performance characteristics the method used has to fulfill. Recognition of this leads directly to the task of the validation of the method one wants to use to get "good" results!

Slide 4
"Whereof one cannot speak, thereof one must be silent" (L. Wittgenstein Tractatus Logicus Philosophicus, 1918). This philosophical phrase of the early 20th century leads us to the problem that we have to define basic analytical terms before we can interpret any experimental data. This slide shows the concept of "precision", i.e. the closeness of agreement between results obtained from measurements of the same sample. As a prerequisite for understanding this concept the students have to realize that due to various reasons (fluctuation of performance characteristics of instruments, slight differences in the chemical status of samples etc.) even the best analytical method/instrument will not deliver exactly the same results for measurements of subsamples of the same "parent sample". The better the closeness of agreement of the obtained results, the better the precision of the analytical method! Or, in the language of analytical chemistry: The precision of a method can be evaluated by performing appropriate series of measurements using distinct subsamples of the same sample. The obtained results reflect the repeatability (same instrument, same observer, same conditions of use, same location, repetition over a short period of time) or the reproducibility (change of instrument, observer, location etc.) of the method. Imprecision is the result of the random distribution of the obtained measurement results around the average value of the analyte.

Slide 5
The term "trueness" also has a philosophical touch because of the underlying ideal concept of the "true value of the analyte" to measure [1]. The true value of an analyte, however, is a goal which cannot be achieved. Still one can use this term as a regulative principle which makes it possible to qualify the *objective* approximation of measurement results to the *truth*. The quantitative expression of this approach is the term "trueness". Trueness is solely influenced by systematic

errors which lead to deviations of the results only in one direction (too low or too high results).

Slide 6
The concept of accuracy is one of the most important pillars of analytical chemistry [2]. It combines the previously described concepts of "trueness" and "precision" and enables the analytical chemist to qualify results using a single term. Ideally, the applied analytical methods are only affected by random errors. In this case "accuracy" and "precision" become equivalent.

Slide 7
Using the newly learned terms and concepts, the students are now able to formulate the following pivotal analytical problem: "How can one be sure that analytical methods produce accurate, i.e. precise and true results?". This question can only be answered by performing a *validation of the method* [2], in which the performance characteristics of the method under investigation are evaluated in order to find out whether a method is fit for the intended analytical purpose. It is obvious that the precision of a method can be evaluated only by using this method properly. The estimation of the trueness is only possible when an independent, already validated method is applied to "control" the results of the method under investigation.

Slide 8
The strategy to validate a method under investigation starts with the selection of an appropriate sample and control method. In case the results obtained from the two methods are in acceptable agreement (see above!) the tested method is fit for the intended purpose.

Slide 9
However, how can this now be demonstrated practically? Students in their first practical course in analytical chemistry usually have to analyse aqueous solutions containing inorganic ions by performing gravimetric and/or titrimetric determinations (the concentration of the solutions are only known by the tutor). The precision of their results can be calculated by the students themselves from the standard deviation, but the trueness of the results they can only prove by application of an *independent comparison method*. "Independent" in this context means: the underlying chemical or physical principle of the control measurement must be different from the one of the standard method applied. Instrumental analysis methods are mostly based on physical principles and therefore are very suitable as comparison methods for methods based on chemical reactions like gravimetry or titrimetry.

Slide 10
Due to its versatility, high sample throughput and robustness flow injection analysis (FIA) is a well suited method to control gravimetric and/or titrimetric determinations performed by students in practical courses. The figure shows the manifold of a simple one-channel FIA-system which allows measurement of typical inorganic ions. The main advantages of the FIA-system developed in this context are its great flexibility, speed and robustness. Measurements can be

carried out in a couple of minutes and when changing the analyte it is only necessary to change one reagent solution and the measurement wavelength - all other parameters remain constant. Only the indirect measurement of sulphate after ion-exchange to nitrate takes more time due to the necessity to obtain a quantitative exchange rate for the ions. Therefore it takes approximately 10 minutes before the solution coming out of the ion-exchange column can be injected into the aqueous carrier stream of the FIA-system.

Slide 11
The determinations of Fe^{3+}, Al^{3+}, PO_4^{3-} and SO_4^{2-} have been selected to demonstrate the validation concept. Solutions containing one of the analytes are handed over to the students in order to perform the analysis first gravimetrically (threefold measurement). A small part of the original solution is used as a subsample to determine the analyte concentration by application of the one-channel FIA-system which working parameters and employed reagent solutions are kept as simple as possible (also threefold measurement). This allows the application of the system even in basic courses.

Slide 12
The linear measurement range of the one-channel FIA-system was adapted to concentrations of Fe^{3+}, Al^{3+}, PO_4^{3-} and SO_4^{2-} typically encountered in basic courses. Analyte concentrations can be detected directly by injection of the sample in the reagent stream. The colour complex formed can be measured using a UV/VIS detector.

Slide 13
The determination of sulphate is different due to the fact that this ion does not form an "easy-to-make" colour complex. Therefore an indirect measurement method was developed: Through an ion exchange column sulphate is converted into nitrate which can be measured directly at 232 nm.

Slide 14
The accuracy of the results achieved using the one-channel FIA-system presented is comparable to the accuracy of gravimetric determinations of typical inorganic ions - assuming proper execution by the students. Therefore the FIA-system is a good choice as an independent comparison method to demonstrate the concept of validation.

Slide 15
This slide is self-explanatory.

Slide 16
Another way of introducing "modern" analytical chemistry into first year practical courses is the implementation of the photometric determination of Cu^{2+} in the form of the $[Cu(NH_3)_4(H_2O)_2]^{2+}$-complex as an independent comparison method to control electrogravimetric determinations performed by the students. The photometric determination shows great simplicity and is therefore well suited for a first year practical course.

Slide 17
Due to the fact that commercial photometers are not designed for didactic purposes, these instruments have a "black box" character which prohibits an in depth view of all relevant pieces of the instrument. To enable the students to get a close look at all the details a modularly constructed photometer was developed which contains only simple and robust components.

Slide 18
In order to build a robust and cheap instrument the optical and detector system was reduced to the necessary minimum. Nevertheless, the arrangement shown in slide 18 allows to perform simple quantitative measurements with sufficient accuracy.

Slide 19 and 20
The modular photometer consists of two cases - one containing all components for the performance of the measurement itself (light source, slit, filter, cuvette holder, detector (photodiode)) and the other containing the necessary electronic components and the display. Both cases are connected with cables for the light source and the photodiode. It is quite obvious that such an "open" instrument allows much more understanding of fundamental measurement principles than a purchasable „black box" apparatus.

Slide 21
The photometric determination of Cu^{2+} in aqueous solutions is a quite simple procedure and has the additional advantage that only inorganic reagents are involved. This facilitates the understanding of the students regarding the formation of the colour complex and the measurement principle.

Slide 22
The accuracy of the results achieved using the" didactic" photometer developed is comparable to the accuracy of the electrogravimetric determination of Cu^{2+} - assuming proper execution by the students. Therefore the photometric method presented here can be used also very well as an independent comparison method to demonstrate the concept of validation.

The examples described show that the implementation of fundamental analytical concepts and instrumental methods into basic courses can be done quite easily and without any question the gain of knowledge for the students is worth the efforts.

References
[1] J. Fleming, B. Neidhart, H. Albus and W. Wegscheider.
    Accred Qual Assur (1996) 1, 135.
[2] J. Fleming, H. Albus, B. Neidhart and W. Wegscheider.
    Accred Qual Assur (1996) 1, 87.

Basic Course Experiments to Demonstrate Validation

(1) Fundamental concepts

(2) Experimental arrangements

(3) Results

Objective of the Lesson

Elaboration of the fundamental meaning of validation:

Validation is the process of making sure that an analytical method is fit for the intended purpose.

or in other words:
Make sure that the obtained results are as good as you need them!

(But first define properly what is „good enough"!)

Fit for Purpose

Question:
How can one achieve good results?

Answer:
by using a method with the appropriate *performance characteristics*
⇩

Task of method validation:

Investigation and evaluation of the *performance characteristics* of an analytical method

Basic Performance Characteristics of Analytical Methods

Precision
or: how good is the closeness of agreement between the obtained measurement results?

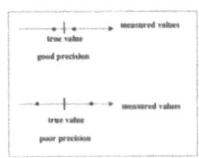

Note:
Precision is only influenced by random errors. This type of error shows no systematics and occurs randomly with statistical variability.

Basic Performance Characteristics of Analytical Methods

Trueness
or: how good is the closeness of agreement between the average value calculated from a series of test results and the (assumed) true value of the analyte (= substance under investigation)?

Note:
Trueness is only influenced by systematic errors. This type of error modifies the result only in one direction (too low or too high results).

Basic Performance Characteristics of Analytical Methods

Accuracy
or: how good is the closeness of agreement between the result of a single measurement and the true value of the analyte?

Note:
Accuracy is a measure which combines precision and trueness (i.e. the effects of random and systematic errors respectively). If the obtained results are not affected by systematic errors, their accuracy becomes equivalent to their precision.

Slides 1-6

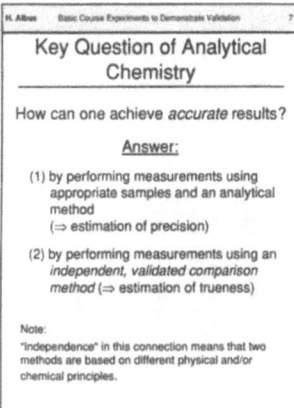
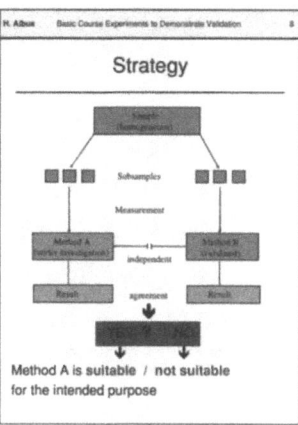
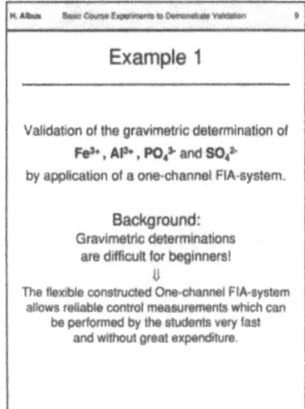
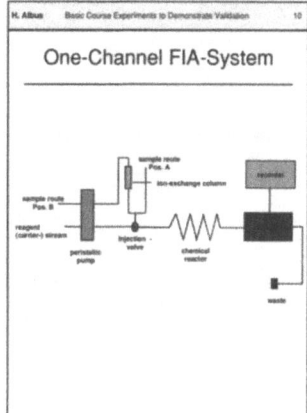
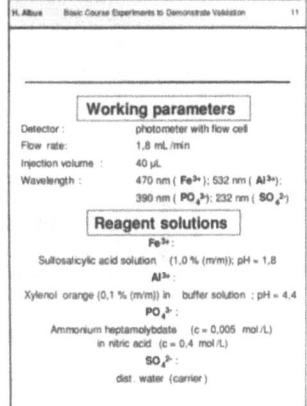
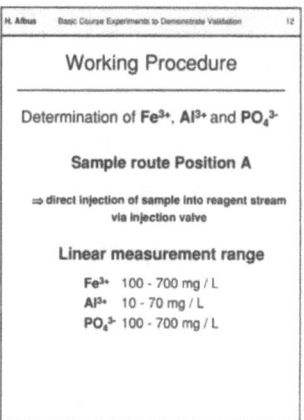

Slides 7-12

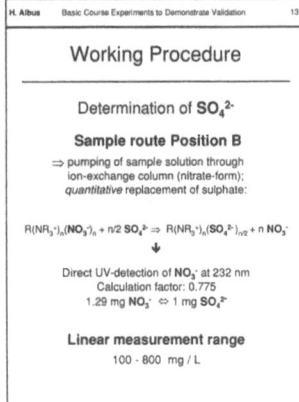
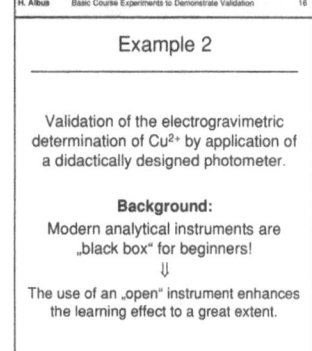
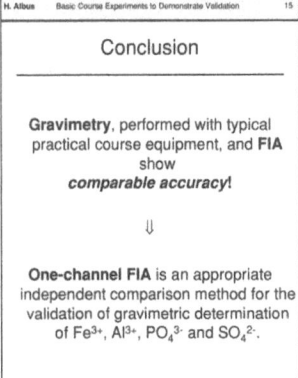
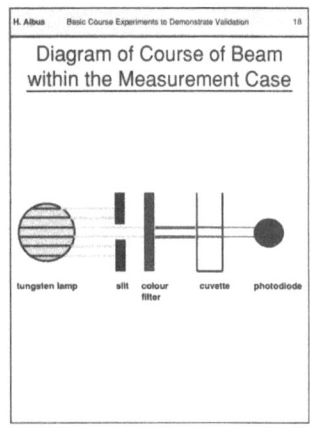

Slides 13-18

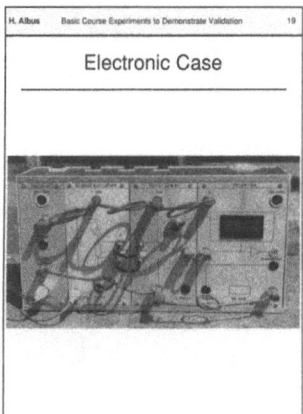

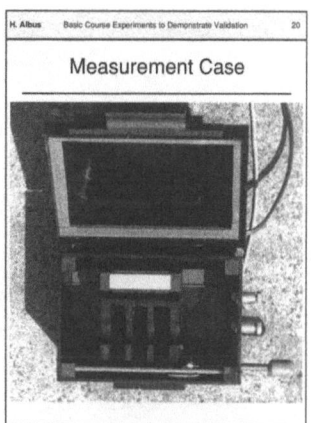

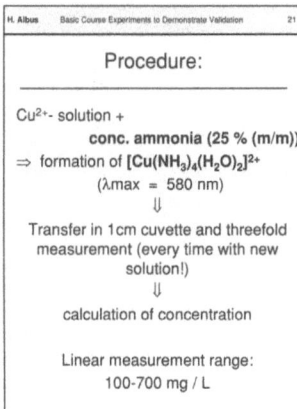

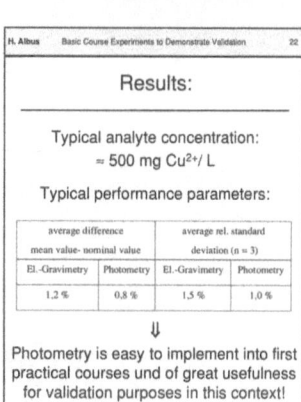

Slides 19-22

Basic Course Experiments to Demonstrate Intercomparisons

U. Pyell

Abstract

Experiments are presented that can be used to demonstrate intercomparisons in a pratical course for first-year students. These experiments include volumetric determinations to be performed by every student, hand-over of data to a supervisor, data evaluation with help of a microcomputer and discussion of underlying statistical principles in a subsequent tutorial. The students learn how to present an analytical result.

At the Philipps-University (Marburg/Germany) undergraduate chemistry students in their first year attend a practical course in inorganic/analytical chemistry. In order to demonstrate the parameters affecting the repeatability and precision of analytical results, a volumetric method is repeatedly applied to aliquots of the same homogeneous sample by every student of this course. Results are compared after statistical treatment of the data by a supervisor.

The sample to be analysed is an aqueous solution of NaOH ($c \approx 0.01 \text{mol L}^{-1}$). Two solutions of the titrator (H_2SO_4) are used ($c_1 = 0.1 \text{mol L}^{-1}$, $c_2 = 0.01 \text{mol L}^{-1}$). Each student repeats the titration three times with each standard solution (six titrations per student). The results are handed out to the supervisor in volume units.

The statistical treatment of the data material includes: verification of Gaussian normal distribution, calculation of arithmetic means, calculation of standard deviations for single measurements and for the mean of three measurements. The results are visualized by diagrams, shown as overhead transparencies.
The presented data is used in a subsequent tutorial:
- to visualize the effects of random and systematic errors
- to introduce statistical methods for evaluation of analytical results
- to discuss the effect of the concentration of the standard solution on the precision of a method
- to introduce a definition of the terms precision and accuracy
- and to explain, why repetition of the method and calculation of the mean enhances the precision of analytical results.

Slide 1

At the Philipps-University Marburg (Germany) undergraduate chemistry students in their first year attend a practical course in inorganic/analytical chemistry. The described experiments are part of the compulsory programme. These experiments include experimental parts that have to be performed by every student (time spent incl. preparation: 2-3 h) and data evaluation performed by a supervisor. The results of both parts are presented to the students via the transparencies in a tutorial (time of lecture: 45-60 min).

Slide 2
The (quantitative) determination that has been selected in order to demonstrate intercomparison of analytical results was the acidimetric determination of sodium hydroxide in aqueous solution with chemical end-point detection (phenolphthalein as visual indicator). Apparatus: 25 mL-pipette, 50-mL-burette, glassware. Two standard solutions (diluted sulfuric acid) are provided: (1) $c(\frac{1}{2} H_2SO_4) = 0.10$mol L^{-1}, (2) $c(\frac{1}{2} H_2SO_4) = 0.0100$mol L^{-1}. Each student analyses 25 mL-aliquots of the same solution containing about 0.01mol L^{-1} NaOH. These concentrations have been selected to demonstrate the impact of a non-appropriate concentration of the standard solution on the precision of a method.

With each standard solution the titration is made three times. Each student has to perform six titrations. The time spent (incl. preparation) does not exceed 3 h. The experiments do neither demand expensive equipment nor skillfull techniques. The results of the series of experiments is handed over to a supervisor. The amount of standard solution that has been consumed per titration is given in volume units. Thus, the result is not biased by calculation errors or rounding.

Slide 3
In a first step the results are grouped into equidistant classes. The distribution of the results can most easily be appreciated by drawing a histogram. This histogram shows that the distribution of the measurements is roughly symmetrical about the mean, with the measurements clustered towards the centre.

Slide 4
In a second step it is investigated, whether the distribution corresponds to a Gaussian normal distribution. To this end, the results are presented in form of a cumulative frequency plot.

Slide 5
In a third step the data have to be normalized to the total number of measurements.

Slide 6
With help of commercial software or of commercial special graph paper (normal probability paper) this curve can be transferred into a normal probability plot. Linearity of the normal probability plot shows that the distribution of data obtained in a series of measurements corresponds to a normal distribution. It must be emphasized that this method can only be used if there are 50 or more data points. Hence, the number of titrations per student has to be adapted to the number of students taking part in the basic course.

The statistical data evaluation is done completely by the supervisor with help of commercial software. The results of the data treatment are presented in form of a tutorial to the students having performed the experimental part.

Slide 7
In this part of the tutorial the students are made familiar with some basic statistical terms. The mathematical description of the probability equation is introduced.

It is shown with the example given that in most cases repeated measurements of a single quantity are normally distributed.

Slide 8
The Gaussian distribution function is introduced, also the parameters μ and σ that characterize this function.

Slide 9
The arithmetic mean $\bar{x}$ as estimate of μ and the standard deviation s as estimate of σ are introduced. It is shown how these quantities can be calculated for a given set of data.

Slide 10
With the example given, it is shown that experimental parameters (here: the concentration of the standard solution) determine the standard deviation for a series of measurements. The relative standard deviation (RSD) is introduced as a measure of the precision of a method. The precision of a method can be improved by variation of experimental parameters (i.e. by adjustment of the concentration of the standard solution to (1) the amount of analyte in the sample and to (2) the dimensions of the burette).

Slide 11
Employing the data material given, the relative standard deviations for the arithmetic means of each three measurements are calculated. Comparison with the RSD for the single results shows that there is some improvement. Hence, the dispersion of the distribution of results can be reduced by formation of the arithmetic mean of the results of repeated measurements.

Slide 12
In the next slide the equation for the standard deviation of the mean is given.

Slide 13
The effect of grouping of data is visualized with a graph depicting normal distribution with different spread.

Slide 14
These considerations should enable students to understand better factors influencing the results of chemical measurents. It is of paramount importance that the difference between random deviations (statistical deviations) and systematic deviations is understood. The term deviation is here preferred to a description of the uncertainty concept, as it is more instructive. All possible deviations that determine the accuracy of the experimental result are discussed with the students and are listed in two groups.

Slide 15 and 16
At this point the terms precision and accuracy can be introduced and explained with help of formal definitions and given examples [1]. In addition slide 13 can be used to visualize the effect of the two types of deviations and their impact on the performance criteria of an analytical method.

Slide 17
The statistical terms confidence interval and confidence limit are introduced. An

equation is given for the calculation of confidence limits. This equation should be used to calculate the confidence interval for a given example (at the blackboard). Knowledge of the meaning of confidence intervals enables the students to make an estimation of the precision of results obtained in further experiments of the basic course. It can be shown easily that repetition of measurements reduces the confidence limit, hence improves the precision of the determination.

If it is intended that analytical results should be presented by the students stating the arithmetic mean of repeated measurements and the confidence limits, a list of t-factors has to be provided as hand-out.

Slide 18
The meaning of the term confidence limit is visualized with help of a graph.

It is important to stress that no quantitative experimental value is of any value unless it is accompanied by an estimate of the uncertainty involved in its measurement.

Slide 19
The possibilities how a quantitative value can be presented are shown: standard deviation or confidence limits. The nonexistence of an universal convention forces to state the form that has been used for the presentation. The number n of repeated measurements has to be given. Provided that the value of n is given, the two forms can be easily interconverted with the equation presented on slide 17.

Slide 20
How analytical results are presented is demonstrated with help of an example. The meaning of the number of significant figures (all digits which are certain plus the first uncertain one) is discussed.

Slide 21
The experimental part and the tutorial should enable students to rationalize precision and accuracy of quantitative measurements. They should understand the difference between statistical and systematic deviations and how they can use statistics to describe the precision of a measurement. They should have a basic knowledge of the Gaussian normal distribution and its parameters. They should be able to present measured data in a reasonable fashion.

References
[1] J. Fleming, H. Albus, B. Neidhart and W. Wegscheider.
Accred Qual Assur (1996) 1, 87.

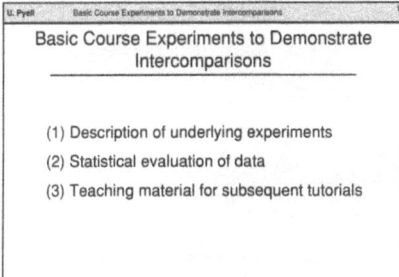
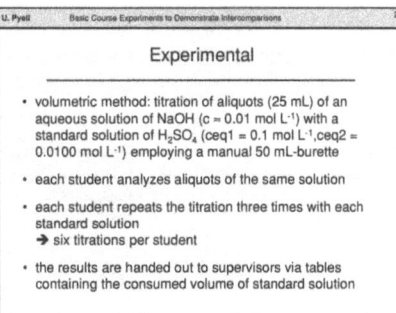
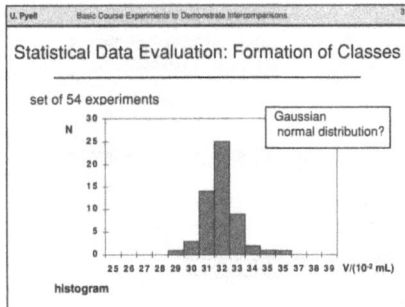
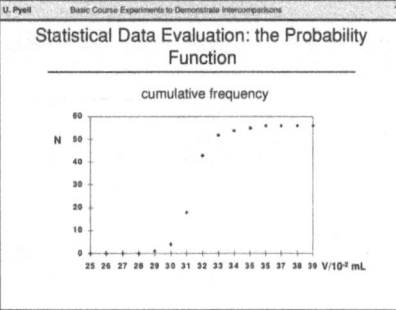
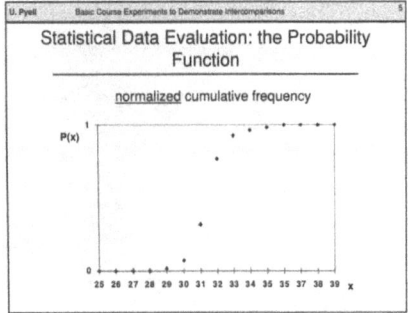
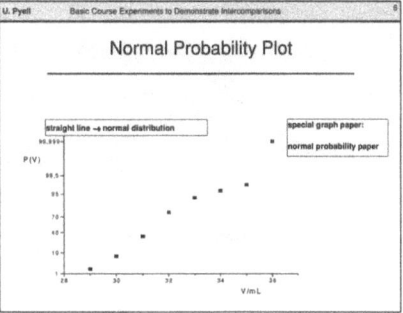

Slides 1-6

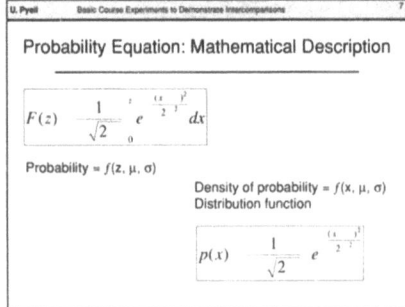
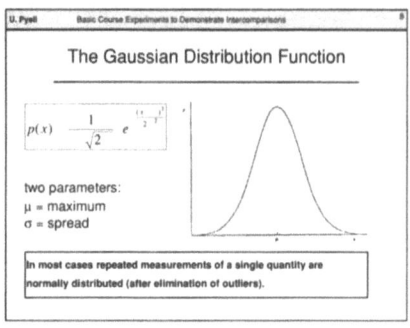
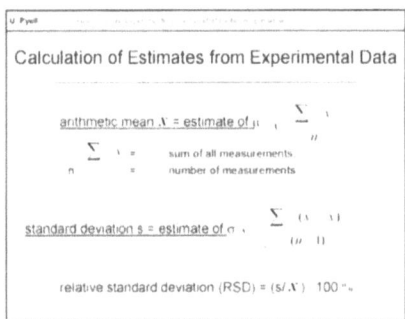
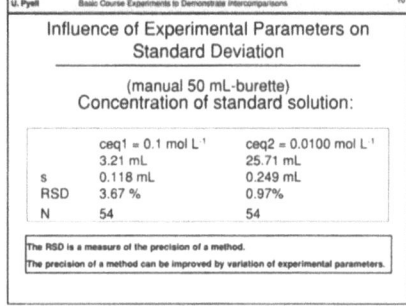
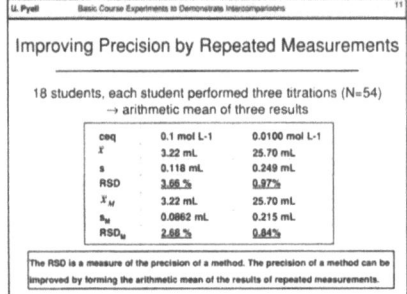
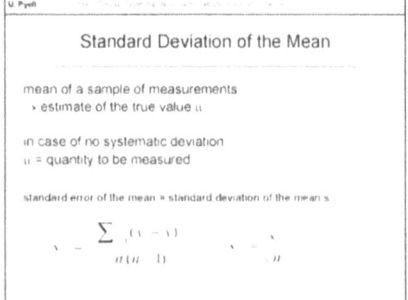

Slides 7-12

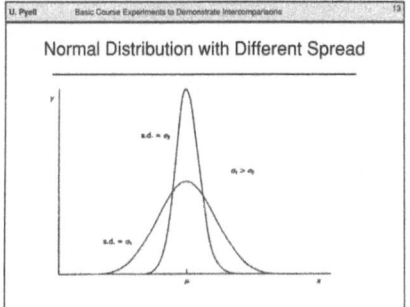
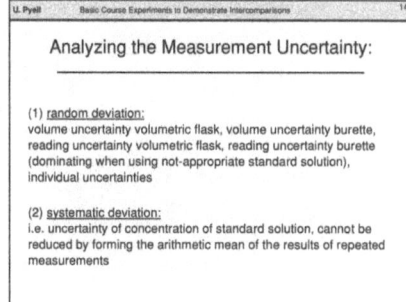
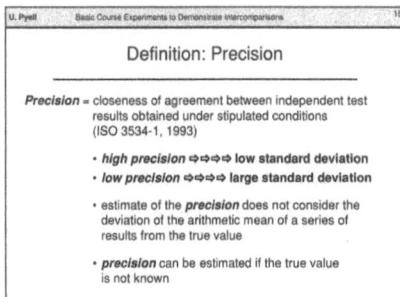
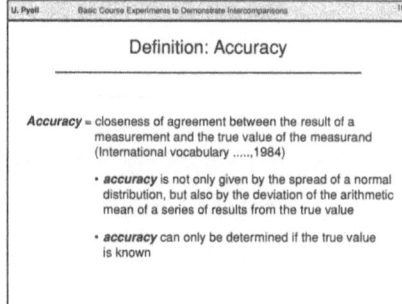
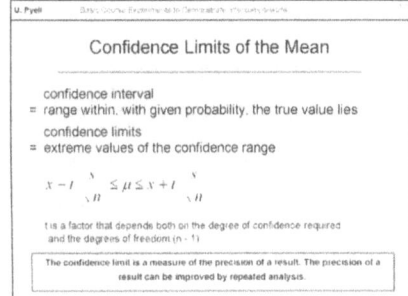
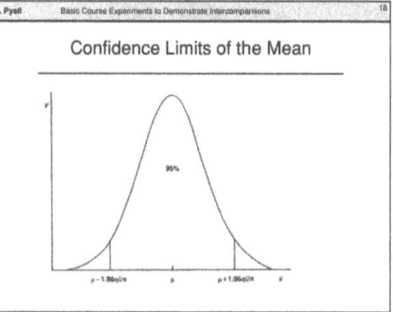

Slides 13-18

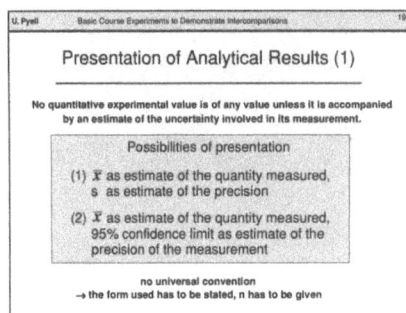
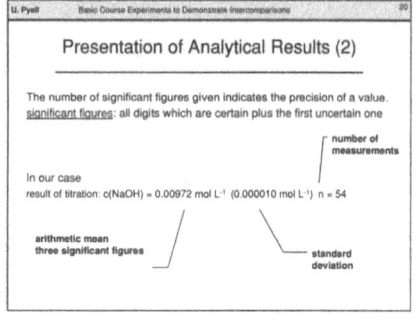
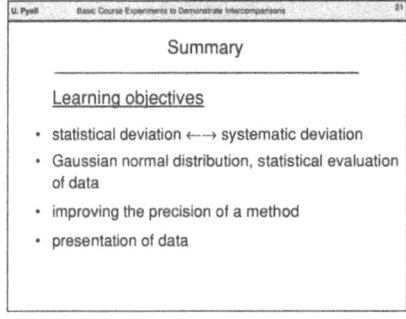

Slides 19-21

Assessment of Test Kits in Terms of Time, Cost and Quality

P. Houlgate

Abstract

The aim of the assignment is to introduce students to the quality assurance aspects of a site survey which need to be considered to ensure client expectations are met. It is designed to enable students to gain experience in formulating a problem, sampling, analysing and reporting. In addition they will also compare laboratory analytical methods and on-site test kits, balancing the three factors of time, cost and quality.

The assignment is to carry out a water survey of a building, comparing use of on-site test kits with traditional methods for measuring specified analytes in the water.

A water quality survey consists of taking a representative number of samples throughout a building and measuring the concentration of an agreed range of analytes to determine whether the levels of analytes measured meet an agreed specification. The cost and time taken to undertake the survey will depend on the purpose of the survey. On-site test kits provide a relatively cheap alternative to traditional sampling and analytical techniques at the expense possibly of quality.

The presentation will describe the components of the assignment which will in general be carried out by small teams of students. The various components will include:
- deciding what analytes to measure in water
- determining the precision of laboratory analytical methods and test kits
- determining the bias of laboratory analytical methods and test kits
- designing a sampling plan suitable for surveying a building
- creating sample information records
- identifying the sampling equipment required to carry out the survey.

Slide 1

The aim of this practical assignment is to compare laboratory analytical methods and on-site test kits when performing a survey of water quality, balancing the three factors of time, cost and quality. The assignment is designed to enable students to gain experience in formulating a problem, sampling, analysing and reporting. Additionally it will introduce students to the quality assurance aspects of a survey which need to be considered to ensure client expectations are met. For the survey to be a success the measurements made must be valid.

Slide 2

The aim of any valid analytical measurement is to deliver reliable measurements first time, every time which will match the needs of the client. To this end a set of six quality assurance principles relating to valid analytical measurement are used by many laboratories in the UK.

The six principles for Valid Analytical Measurement (VAM) are
- Analytical measurements should be made to satisfy an agreed requirement.
- Analytical measurements should be made using methods and equipment which have been tested to ensure they are fit for purpose.
- Staff making analytical measurements should be qualified and competent to undertake the task.
- There should be a regular independent assessment of the technical performance of a laboratory.
- Analytical measurements made in one location should be consistent with those elsewhere.
- Organisations making analytical measurements should have well defined quality control and quality assurance procedures. The aim of the VAM programme is to improve analytical measurements in the UK.

The VAM principles were formulated by a group of analytical scientists working on quality issues as part of the Valid Analytical Measurement (VAM) programme which is funded by the Department of Trade and Industry (part of the United Kingdom Government). The VAM programme is part of the National Measurement System so sits alongside standards of length, mass and time. The VAM principles are not a quality system, more a way of developing good laboratory practice. The implementation of the VAM principles is facilitated with a growing range of VAM products.

Slide 3

A client has rung the laboratory where the student is working asking if a water quality survey can be undertaken. Obviously this information in insufficient and the first task the student has, is to find out more about the client's requirements. This ties in with the first VAM principle about the need for an agreed analytical requirement. N.B. Client's invariably want surveys which are cheap, accurate and can be delivered in a very short time period!. The student will need to determine whether using on-site test kits or laboratory analytical methods will be most suited to meet the client's requirements.

The scenario given here is very general to allow for a wide range of options. You may want to amend the scenario so that it can be more focused to meet your needs and the needs of your students.

Slide 4

There are four typical reasons as to why the client requires a survey:

Regulatory requirement: The client may need to know that the water quality complies with a regulatory requirement. A typical example is that the client needs to know that the quality of the drinking water supply throughout a large building meets the appropriate regulations.

Monitoring: The client may require a regular monitoring programme to be able to identify problems at an early stage and carry out remedial action to minimise the chances of costly repairs at a later date. For example, the client may be responsible for a large site with several old buildings. The client needs a regular survey to identify where corrosion is taking place in the hot water system so that effective remedial action can be taken at an early stage.

Identification: The client may have a problem and requires a survey to help identify the cause of the problem. For instance a child might have become ill with stomach pains and been rushed to hospital when swimming at the local pool. The client wants to check that the swimming pool water is being chlorinated effectively and so eliminates the possibility of it being a bacteriological problem.

Dispute: There may be a disagreement between the client and the supplier which requires a survey to resolve the disagreement. For instance, a manufacturer may be in dispute with a trading standards officer as to whether the description on the outside of a bottle of sparkling mineral water is a true representation of the contents. The manufacturer has labelled the contents of the bottle as 'benzene free' whereas the trading standards officer suspects that there might be traces of benzene present because of problems with the carbonation process.

Specific recommendations have not been made about what should be measured. The choice of what to measure in the survey will depend on
the age and aptitude of the student,
the laboratory equipment available,
the time available to complete the assignment,
the cost of the test kits and
the need for students to develop analytical skills in particular areas.

It is envisaged that at one end of the spectrum you may have a young student taking a couple of samples of swimming pool water and determining pH, sulphate and chlorine using on site test kits and classical analytical methods. At the other end of the spectrum, a student may take several samples to determine the extent of corrosion in the hot water supply. Analytical technique such as GFAAS or ICP - OES would be used and the results compared to those from on site test kits to determine which would be most suitable to meet the clients requirements of cost, time and quality. Quality in terms of reagent blanks, sample blanks and QC samples for monitoring purposes might be part of the latter project.

Slide 5
Students will perform a survey of water quality to
- satisfy the agreed client's requirements and
- compare the performance of test kits with laboratory methods in terms of cost, time and quality and
- learn how the VAM Principles provide a framework to achieve this objective.

At this stage the tutor will describe the type of survey the students are to perform and the analytes to be measured.
Four possible types of survey envisaged are:
site survey - drinking water,
site survey - corrosion survey,
survey of swimming pool water,
survey of mineral water. It may be more suitable to choose a different type of survey to meet the needs of the students, e.g. a survey of river water quality.

In each case the students will have to (a) design a sampling plan, (b) evaluate the use of on site test kits and laboratory methods, (c) undertake the sampling survey (d) perform the analysis, (e) carry out statistical analysis of the data and (f) write a report of their findings.

N.B. A site survey of water quality consists of taking a representative number of samples throughout a building and measuring the concentration of an agreed range of analytes to determine whether the levels of analytes measured meet an agreed specification. The cost and time taken to undertake the survey will depend on the purpose of the survey. On-site test kits provide a relatively cheap alternative to traditional sampling and analytical techniques though possibly at the expense of quality.

Slide 6
An example is given here concerning pH measurement and what method evaluation is required. This example has been selected because students are familiar with both the on-site test kit (pH paper) and the laboratory analytical method (glass pH electrode). What are the client's requirements when measuring pH? The answer depends on why the measurement is being performed. Swimming pool water must have a pH between 7.2 and 7.8 (and ideally better than that) to allow effective chlorination of the water to take place. Drinking water has a wider range (5.5 to 9.5), selected to prevent significant corrosion occurring in the supply system. How are students going to determine the criteria for deciding which method is most suitable in terms of trueness, cost and precision?

Tutors may want to select a different example for their students here. If necessary the concept 'method validation' can be explained at this point.

Slide 7
One of the main aims of the survey is to compare the use of on-site test kits with laboratory analytical methods in terms of time, cost and quality. Any evaluation of the performance of on-site test kits and laboratory methods in terms of quality needs to answer these questions:
- Are the test kits/lab methods sufficiently accurate?
- Are the test kits/lab methods sufficiently precise?

The pH measurement described in the last slide is a good example of what is meant by fit for purpose in terms of quality. For many applications a piece of pH paper is sufficiently precise and accurate to measure the pH of a sample. This is because the client only requires the pH to be measured to the nearest integer (e.g. determining the pH of garden soil). Occasionally, the pH needs to be determined more precisely and accurately (e.g. in the case of swimming pool) in which case the pH paper might be unsatisfactory.

Remember there is no point in doing a measurement if you cannot achieve satisfactory quality to meet the client's requirement. Either a more suitable analytical method must be found or discuss the possibility of modifying the requirement and developing an alternative strategy with the client.

Are the test kits/lab methods under control?
The fact that the method performs satisfactorily at the time when it is validated is not sufficient. Analysts require analytical methods to perform reliably every time. Students will be introduced to the use of quality control charts to demonstrate that methods are under control.

What other factors should be taken into consideration when deciding that a method is 'fit for purpose' in terms of quality?

Slide 8
VAM principle 2 states that "Analytical measurements should be made using methods and equipment which have been tested to ensure they are fit for purpose". Depending on the clients requirements, various method parameters might need to be checked to determine if the method is fit for purpose. For example, a gravimetric method is not used to measure the lead concentration in drinking water because it cannot detect lead at the $\mu g\ dm^{-3}$ level.

Method validation parameters to be tested might include:
- selectivity
- range
- linearity
- limit of detection and quantitation
- ruggedness
- precision and bias (uncertainty).

It is suggested that the tutor determines which aspects of method validation the student needs to undertake as part of the assignment. The students knowledge will determine how much explanation is required.

Slide 9

Part of this assignment is to compare the use of on-site test kits with laboratory analytical methods in terms of cost and time. However, what is meant by cost and time?

Can the client afford the cost of the analysis?

The pH paper has the advantage in cost (i.e. it is a lot cheaper to purchase) compared to the laboratory method. Generally there is no point undertaking an analysis if the client does not have the money to pay for it. In this assignment, cost is defined as the expense of purchasing the equipment and the number of 'man hours' to perform the measurements.

Can the laboratory produce the results in time to meet the client's requirements?

The pH paper has the advantage of time, you can perform the measurement immediately. With the laboratory method, you have to take a sample and transport it to where the equipment is before undertaking the measurement. This can be a serious disadvantage when the pH of the sample changes significantly with time. In this assignment, time is defined as the period (in days) between taking the samples and producing a written report of the results to give to the client.

Is the method fit for purpose?

All of these factors (time, cost and quality) are used to determine if a method is 'fit for purpose'. In other words, a method is said to be 'fit for purpose' if it meets the client's requirements of time, cost and quality. In practice laboratories often sacrifice time and cost to ensure that the results are of satisfactory quality.

There is another factor to be taken into consideration when undertaking a task for the client.

The concept of what is involved in terms of cost and time has been radically simplified as the main aim of the assignment is to get the students to think about the quality aspects of the survey.

Slide 10

VAM principle 3 states that "Staff making analytical measurements should be qualified and competent to undertake the task."

It is sometimes forgotten that validating a method by itself is not enough. If one wants good results then one needs to ensure that the staff undertaking the work are qualified, trained, motivated and rewarded financially. Also the procedure for performing the method must be documented and written in a language that can be understood by the analyst.

For this reason it is suggested that students undertake the validation experiments on precision and bias before carrying out the survey. This will enable students to become trained in the use of the test kits and the laboratory methods. It should show up any problems that the student has with getting the methods to perform satisfactorily.

Slide 11

One of the key method validation parameters is precision. There is little point in making a measurement if the next time the measurement is performed on an identical sample, a different result is obtained.

A large homogeneous sample is needed to validate the method for precision. In addition the sample must contain a measurable amount of the analyte under investigation, ideally at a concentration close to that expected in the actual samples. Six aliquots from the large sample are taken and analysed. The results can be used to calculate an overall mean, standard deviation, coefficient of variation and 95% confidence limits. This information can be used to make a judgement about whether the method is precise enough to meet the client requirement's. This set of experiments will also provide the student with information about the time taken to undertake the analysis.

The students should be given sufficient information so that they can perform a set of repeat measurements on a homogeneous sample and be able to carry out simple statistical analysis on the analytical data to determine the precision of the technique. If appropriate the topics of reference materials, blanks and spiked samples could be introduced here. The tutor may want to discuss the difference between precision and bias at this point.

Slide 12

VAM principle 6 states that "Organisations making analytical measurements should have well defined quality control and quality assurance procedures."

One such procedure is the use of a quality control chart. The analytical method(s) should be able to produce the same results when used to analyse identical samples on different days. To ensure that the method is under 'statistical' control, a quality control sample (QC sample) is measured regularly and recorded systematically. The homogeneous sample used for making the precision measurements (if stable) is ideal to use as a QC sample and the results from the precision measurements can be used to prepare the chart.

This section is to introduce students to the role of quality control charts. Tutor's can leave out this section if they want. Tutor's may need to explain about the action and warning limits on the control chart.

Slide 13

The students should be given sufficient information so that they can perform measurements on homogeneous samples of known concentration and be able to carry out simple statistical analysis to determine the bias of the technique. The topic of various types of standards, reference materials and spiked materials could be introduced here. N.B. Three analyses at a particular concentration is sufficient if the precision of the measurement has been determined.

Another key method validation parameter is bias. In an ideal world there should be no bias in the results.

A homogeneous sample is needed to validate the method for bias. In addition the sample must contain a known amount of the analyte under investigation, possibly at a concentration close to that expected in the actual samples or at concentration close to a specified limit. Three aliquots from the sample are taken and analysed. The results can be used to calculate an overall mean and standard deviation. The student will compare the mean with the known concentration to determine the bias for the method. The standard deviation can be compared with that calculated from the precision experiments to show that the method is under control.

Slide 14

The students will probably need to perform other measurements to validate the method. For example, the student should check the linearity and range of the calibration when making spectrophotometric measurements. Slides for other method validation measurements should be prepared in a similar fashion to slide 12.

The aim of the method validation measurements was to ensure that the methods/test kits are 'fit for purpose' in terms of quality. The student should have enough information to be able to make a judgement about the following questions:

- Are the methods sufficiently accurate and precise for their intended use?
- Do the methods give the same results with identical samples on different days?
- Do the methods perform satisfactorily in terms of the method parameters checked (e.g. limit of detection)?
- Are the methods cost effective?
- Can the results be produced within a reasonable time frame?

Slide 15

The examples given in this section relate to a survey of a swimming pool. The examples should be modified if the students are performing a different type of survey.

The analytical measurements are only part of the water quality survey. There is also the need to take appropriate samples. The student needs to devise a sampling strategy to meet the needs of the client. The student should consider the following points when developing a sampling strategy:

- Determine the number of samples to be taken.

  For example, how many samples would be representative of the whole swimming pool? How many samples is it practicable to take in the time available allowing for the fact that on site measurements have to be performed as well?

- Identify where the samples are to be taken from.

  For example, should the samples be taken from the shallow and the deep end? Should the samples be taken from near the surface of the pool? Should a sample be taken from the point where treated water enters the pool?

- Determine what size of sample to take.

  For example, have you taken a sufficiently large sample for all the analyses? What equipment have you available for storing the sample?

- Decide if the samples need special storage conditions.

  For example, samples for chlorine analysis should be refrigerated as soon as practicable, samples for metal analysis should be collected in plastic bottles and preserved by the addition of a small amount of nitric acid.

Slide 16

It is important to make "sample observations" and record relevant information. This information will be used in the interpretation of the results of the analysis and also to make judgements regarding the time taken to undertake the survey and on site measurements.

The student should design a form to record the following information:
- purpose for taking sample (e.g. to identify if the swimming pool water should be replaced)
- identification numbers and traceability of samples (trackability)
- sample location and date of sampling
- sample appearance (e.g. clear, colourless, trace brown sediment, faint chlorine odour)
- observations (e.g. brown stains on tiles of swimming pool).

N.B. Results of the on site measurements and the time taken to perform the sampling and analytical measurements could also be recorded on these forms.

Slide 17

A good way to prepare for the sampling operation would be to undertake a small practice survey first. Students could practice at home taking a sample of bath water and carrying out on site measurements.

The student needs to consider the following factors when performing the survey:
- Can the survey be performed safely (e.g. the student has a safe way to carry glass bottles)?
- Has the student got the correct equipment for taking samples (1 dm^3 glass bottles are suitable for most analytes)?
- Has the student thought to take pens, paper, rubbish bags, etc.?
- Does the student know how to maintain sample integrity until analysis can be performed?

Slide 18

The day has arrived for undertaking the survey!
It is worthwhile the student sitting down for five minutes and checking that the points listed have been checked and completed:
- devised a sampling strategy
- created a system for recording sample information and on site test results
- made suitable preparations for the sampling operation
- know how to use the on site test kits
- planned how to preserve the integrity of the samples.

Slide 19

Performing the analysis should not present any major problems. The student should have gained sufficient experience in using the analytical methods and test kits when carrying out the method validation. The analysis part of the survey will be a success if you have:
- checked the methods perform satisfactorily
- been trained in the use of the analytical methods and test kits
- in place a regular scheme for regular monitoring of the performance of the laboratory analytical methods and on site test kits
- relevant sample information to link to the analytical results.

For example, brown stains observed in a hand basin and brown sediment present in the sample of water would tie in with an analytical result showing that there is a high concentration of iron in the sample.

Slide 20
The tutor may want to be very specific about what is required in the report. This will enable the tutor to make a valid assessment of the student's performance.

Look again at VAM principle 1:

Analytical measurements should be made to satisfy an agreed requirement.

Writing a report to a client is an essential part of this principle. If you fail to communicate effectively the results of your findings to the client then this principle has been breached as the client is not satisfied. Beware of writing reports in a language that the client can not understand. Beware of calculation or transcription errors. They are two of the most common reasons why the client is given the wrong result.

There are two types of report that can be written depending on the client's requirements:
- A report to a client detailing the findings of the survey regarding the quality of the water.
- A technical report to an expert regarding the use of test kits in terms of cost, time and quality.

Slide 21
The aim of this practical assignment was to compare laboratory analytical methods and on-site test kits when performing a survey of water quality, balancing the three factors of time, cost and quality. To perform this survey satisfactorily you have used four of the Valid Analytical Measurement principles. They are:
- Analytical measurements should be made to satisfy an agreed requirement.
- Analytical measurements should be made using methods and equipment which have been tested to ensure they are fit for purpose.
- Staff making analytical measurements should be qualified and competent to undertake the task.
- Analytical measurements made in one location should be consistent with those elsewhere.

Organisations making analytical measurements should have well defined quality control and quality assurance procedures.

Quality Assurance principles such as the VAM principles apply to all aspects of the task and is something students should use whatever field of science they work in.

A Survey of Water Quality

Comparison of on-site test kits and laboratory analytical methods when performing a survey of water quality balancing the three factors of time, cost and quality.

A successful survey requires valid analytical measurements.

Six Principles for Valid Analytical Measurement

- Agreed analytical requirement
- Methods/equipment fit-for-purpose
- Competent staff
- Regular checks on technical performance
- Consistency of measurements
- Quality Assurance/Quality Control

The Scenario

You have received a phone call from a client asking if you can perform a water quality survey.

Possible Client Requirements

Why does the client require the survey?

- Regulatory requirement (e.g. drinking water quality fit for human consumption)
- Monitoring (e.g. survey for corrosion within the hot water supply)
- Identification (e.g. people taken ill in a swimming pool)
- Dispute (e.g. is there benzene in mineral water)

The Student's Assignment

The assignment is to :

- perform a survey to satisfy the client's requirements.
- compare the performance of test kits and laboratory methods in terms of time, cost and quality.
- learn how the VAM Principles provide a framework to achieve this objective.

Comparing the Performance of Methods: e.g. pH

Client's requirement	Test kit method	Laboratory method
	Universal pH paper range 0-14 in steps of 0.5	Calibrated pH electrode
Swimming pool water pH range: 7.2 and 7.8 *ideally 7.4 and 7.6*	trueness (bias)?	
	cost?	
Drinking water supply pH range:5.5 and 9.5 *ideally 6.5 and 8.5*	precision?	

Slides 1-6

Performance in Terms of Quality

- Are the test kits/lab methods sufficiently accurate?
- Are the test kits/lab methods sufficiently precise?
- Are the test kits/lab methods under control?

Method Validation - Parameters

Method validation might include:
- Selectivity and specificity
- Range
- Linearity
- Limit of detection and limit of quantitation
- Ruggedness
- Precision and bias (uncertainty)

Performance in Terms of Cost and Time

- Can the client afford the cost of the analysis?
- Can the laboratory produce the results in time?
- Is the test kit/lab method 'fit for purpose'?

Competent Personnel - Staff Validation?

People who make analytical measurements should be:
- Qualified
- Trained
- Motivated
- Rewarded financially

Precision

For precision measurements you will need:
- a large homogenous sample.
- the analyte to be present in the sample matrix at an appropriate concentration.
- to perform at least six repeat measurements.
- to perform a statistical analysis (mean, standard deviation, coefficient of variation and 95% confidence limits).
- to use the statistical analysis to make a judgement.

The Role of Quality Control Charts

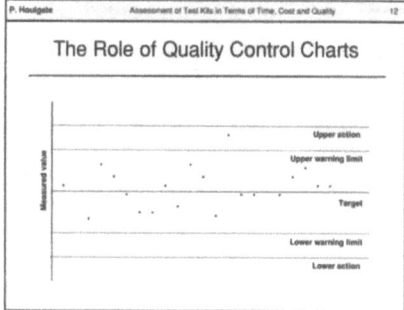

Slides 7-12

Bias

For bias measurements you will need:
- a homogenous sample containing a known concentration of analyte (e.g. a reference material).
- to perform at least three repeat measurements.
- to perform a statistical analysis (mean, standard deviation).
- to use the statistical results to make a judgement about bias.

Are the Methods / Test Kits "Fit for Purpose"

- Are the methods accurate and precise enough for their intended use?
- Do the methods perform reliably (i.e. reproducible results)?
- Do the methods perform satisfactorily (e.g. limit of detection)?
- Are they cost effective?
- Can they produce the results in time?

Sampling Strategy

Use the client's requirements to:
- determine the number of samples to be taken.
- identify where are the samples to be taken from.
- decide what size of sample is required.
- decide if the sample needs special storage conditions.

Sample Information Records

- Purpose for taking sample
- Identification numbers & traceability of samples (trackability)
- Sample location and date of sampling
- Sampling appearance
- Observations

Preparing for the Sampling Operation

- Carry out a trial run
- Think about safety (e.g. carrying the sampling equipment).
- Identify and obtain appropriate sampling equipment.
- Make a list of everything you require to complete the survey.
- Plan how you are going to maintain sample integrity.

Undertaking the Survey

The survey will be a success if you have:
- devised a sampling strategy.
- created a system for recording sample information and on site test results.
- made suitable preparations for the sampling operation.
- learnt how to use the on site test kits.
- made arrangements to preserve the integrity of the samples.

Slides 13-18

Undertaking the Analysis

The survey will be a success if you have:

- checked the methods perform satisfactorily.
- been trained in the use of the methods.
- in place a quality control scheme for regular monitoring.
- relevant sample information to link to the analytical results.

Writing the Report

There are two types of report that can be written depending on the client's requirements:

- A report to a client detailing the findings of the survey regarding the quality of water.
- A technical report to an expert regarding the use of test kits in terms of cost, time and quality.

The Role of VAM in Performing the Survey:

- Agreed analytical requirement
- Methods/equipment fit-for-purpose
- Competent staff
- Consistency of measurements
- Quality Assurance/Quality Control

VAM principles apply to all aspects of the survey

Slides 19-21

Estimation of Random Deviations in Analytical Methods using Analysis of Variance

E. H. Evans

Abstract
Analytical methods can vary from extremely simple one-step procedures to those which include a number of steps, often involving complex sample treatments and manipulations, and it can often prove extremely difficult to isolate the variability associated with each step in the method. Proper experimental design can yield more information than an ad hoc approach, and subsequent statistical analysis can reveal the magnitude of random deviations in the steps of a multi-step method, which are not always obvious during a cursory examination. In this exercise, Analysis of Variance (ANOVA) is used to estimate the variance components in a multi-step analytical procedure, namely a sequential extraction of trace elements from soil, with subsequent analysis by flame atomic absorption spectrometry. The exercise will be used to demonstrate the following:
- the principles of good experimental design
- the importance of replication and randomisation
- the use of F-tests and hypothesis testing
- the estimation of random deviations in an analytical method.

The exercise is aimed at final year undergraduate students, or postgraduate students. The students are required to perform the analytical procedure as part of a laboratory exercise, working in a group and pooling the data at the end of the lab. A subsequent tutorial will then be used to analyse the data and draw conclusions about the analytical method.

Slide 1
In this exercise, Analysis of Variance (ANOVA) is used to estimate the variance components in a multi-step analytical procedure, namely a sequential extraction of trace elements from soil, with subsequent analysis by flame atomic absorption spectrometry. The exercise will be used to demonstrate the following:
- the application of good experimental design
- the importance of replication and randomisation
- the use of F-tests and hypothesis testing
- the estimation of random deviations in an analytical method.

Slide 2
Experimental design can be used to yield more information about a procedure than an ad hoc approach. Most experimental design addresses the effects of different variables or *factors* operating on a response variable. These factors can be at different *random* or *fixed* levels.

For example, in a study of the effect of temperature on reaction rate, the factor would be temperature which could be set at randomly chosen levels. The response variable would be reaction rate.

Provided that the principles of good experimental design are adhered to, and replication and randomisation is included in the procedure, statistical techniques such as Analysis of Variance (ANOVA) can be used to estimate the variance components, and F-tests applied to determine whether the variance of any step in the procedure is entirely due to random deviations.

Slide 3
Most analytical methods include a number of steps which follow one after another in a logical order. In this case the factors are said to be *hierarchical* or *nested*. In such methods it is often difficult to deconvolute the relative variance contributions of each step from the overall variance.

Slide 4
In this exercise the sequential extraction of Zn from sediment is used as an example of such a method. A schematic of the method is shown here. As can be seen, it includes a number of steps which are hierarchical, and errors could be introduced at each stage which would influence the precision of subsequent stages. In order to estimate these errors *replication* must be introduced at each stage of the method.

Slide 5
The inclusion of replication results in the experimental design is shown. In this case, three samples of the original sediment are taken and subjected to the first extraction step, then three sub-samples are again taken and subjected to the second extraction step. Finally, the concentration of Zn is determined in the final extractant solution by flame atomic absorption spectrometry..

Slide 6
The concentrations of Zn determined in each of the solutions are the response variables, i.e. the measured or dependent variables, the magnitude of which are determined by the factors operating on them. Hence, the response variable is the sum of the true value (μ), plus a function of the factors (τ) plus a function of the overall random deviations (ε). The subscripts i,j,k simply denote a particular combination of factor levels at which the response has been measured. In the case described here, the statistical model for the method is given in Eqn. 1:

$$Y_{ijk} = \bar{x} + E1_i + E2_{j(i)} + Error_{k(ij)} \tag{1}$$

Where the variables *E1* and *E2* are the main effects for factors *Extraction1* and *Extraction2* respectively; and the true value has been replaced with the mean. In this particular case the factors are nested, hence the subscript *j(i)*, for factor *E2*, *and there are no interaction effects because of this*.
A simpler way to express the model is given in Eqn. 2:

$$Zn\ conc. = \bar{x} + Extraction1 + Extraction2(Extraction1) + Error \tag{2}$$

Slide 7
In a similar way the overall variance of the individual measurements can be broken down into the variance components associated with the individual factors:

$$s^2_{Total} = s^2_{Extraction1} + s^2_{Extraction2} + s^2_{Error} \qquad (3)$$

Where S is the standard deviation, and S^2 the variance. In practice the variances corresponding to the individual factors can be estimated by solving the expressions for the estimated mean squares.

Slide 8
The only hypothesis that can be tested is the null hypothesis (H_0). In this case the null hypothesis is: "The extraction does not contribute significantly to the overall variance of the method".

Slide 9
It is important to note that one can only test hypotheses on a probability basis, hence, by utlizing the appropriate F-test and probability tables, the probability of rejecting the null hypothesis when it is true can be obtained. If P = the probability of rejecting H_0 when it is true then:

$P < 0.001$: very strong evidence against H_0 (i.e. 99.9% confidence level)
$P < 0.01$: strong evidence against H_0 (i.e. 99% confidence level)
$P < 0.05$: some evidence against H_0 (i.e. 95% confidence level)
$P > 0.05$: little or no evidence against H_0.

The $P < 0.05$ value, or 95% confidence level, is often taken as the basis for making decisions on significance, though higher levels of confidence might be more desirable.

Slide 10
The analysis will be performed on the nine solutions resulting from the 2nd extraction step. It is necessary to make three separate determinations of the concentration of Zn in each of these nine solutions, thereby resulting in 27 measurements in total. The order of analysis is randomised by writing identifiers for each of the 27 determinations on a piece of paper and drawing them out of a bag in a random manner. An example of such a random order is shown here.

Slide 11
Analysis of the extracts results in the kind of data shown here for the concentration of Zn. The data should be ordered logically as shown and subject to ANOVA.

Slide 12
Results of the ANOVA
In this case, factors E1 and E2(E1) have $P>0.05$ and $P<0.001$ respectively, so there is very little probability of rejecting the null hypothesis when it is true with respect to factor E2(E1). In other words, there seems to be a large relative contribution to variance caused by the differences between samples after the 2nd extraction but not the 1st. In terms of the usefulness of the method of sequential extraction this tells us that any variance that was introduced into the results by the 1st extraction did not have a significant effect on the 2nd extraction.

Estimation of Random Deviations in Analytical Methods using Analysis of Variance

Objectives
- Application of good experimental design
- Importance of replication and randomisation
- Use of F-tests and hypothesis testing
- Estimation of random deviations in an analytical method

Experimental Design

- Fixed and random effects
- Factors and levels
- Replication
- Randomisation

Analytical Methods

- Can be simple one-step procedures or involving multiple steps and complex manipulations
- Tend to be hierarchical in nature
- Difficult to estimate the relative variance contribution of each step

Schematic of Hierarchical Extraction Method

Extract with acetic acid for 30 min. Centrifuge and separate supernatant

Extract with hydroxylammonium chloride for 30 min

Analyse supernatant using FAAS

Experimental Design for Extraction of Zn

1st Extraction
2nd Extraction
Replicate analyses

The Statistical Model

$$Y_{ijk} = \bar{x} + E1_i + E2_{j(i)} + Error_{k(ij)}$$

Zn Conc.= x + Extraction1 + Extraction2(Extraction1) + Error

Slides 1-6

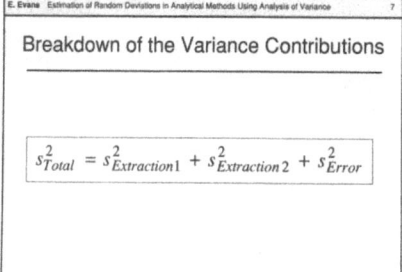
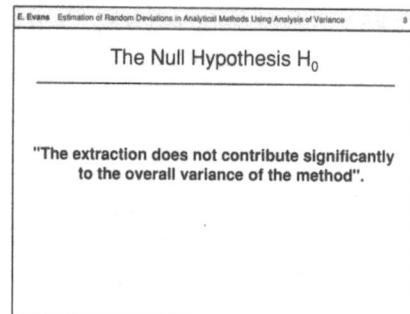
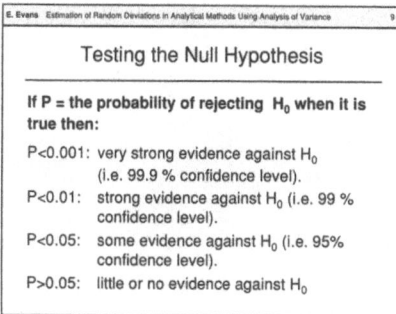
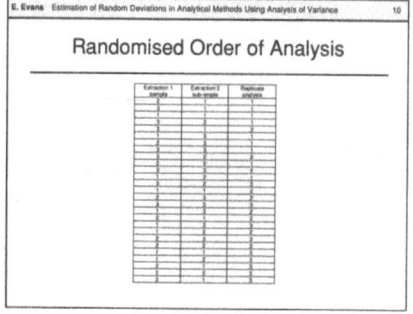
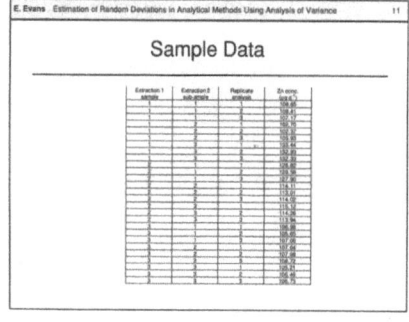
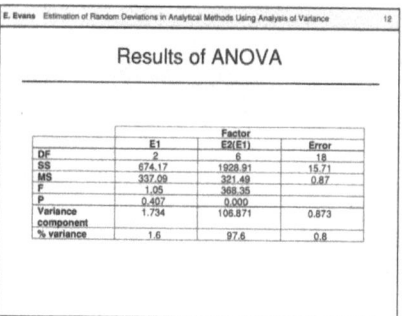

Slides 7-12

Course Structures, Contents and Experiences

PT Scheme for Pre-university Students

E. Lee

Abstract

As part of the Valid Analytical Measurement (VAM) programme and in conjunction with Nuffield Science, an organisation involved with vocational and academic qualifications, LGC has developed a proficiency testing (PT) scheme appropriate for pre-university students.

All PT schemes share two key objectives:
- the provision of a regular, objective and independent assessment of the accuracy of analytical laboratory's results on routine test samples
- the promotion of improvements in the quality (accuracy) of routine analytical data.

In broad terms these are the objectives of the PT scheme organised by LGC for schools and colleges. The students, working in groups, have to determine the concentration of ethanoic acid samples using good laboratory practice. The PT scheme is conducted in a similar manner to PT schemes used by professional analytical chemists and as such

- vialed samples of ethanoic acid are distributed to all participating centres
- the homogeneity and stability of the acid solutions are measured by analysts at LGC
- statistical analysis is performed on the students experimental data and their calculations of the acid concentrations.

In this way, the performance of students in one centre is compared with that of students at the other centres and this information is relayed back to the participating centres.

The PT scheme has been run on two occasions and from teacher's feedback, the PT scheme is a 'user friendly' activity for both students and teachers. Participating centres also receive teachers notes, student information sheets (practical details) and student reporting sheets. In the light of teachers comments, the difficulties encountered by students performing the analysis and the student reports a 'Guide to Improving Analytical Quality in Chemistry' has been produced to help students with analytical techniques such as titration.

The PT scheme for schools and colleges runs over an entire school term (approximately 3 months) and so this allows students to perform the analysis when it best fits in with their studies. It provides a friendly analytical challenge for students and, at the same time, it has helped to improve the way in which students perform their measurements.

Slide 1

The Valid Analytical Measurement (VAM) programme is part of the UK National Measurement System (NMS). The Education and Training component of VAM has one section specifically dealing with pre - university studies. Initially the aim of this component was to provide support materials for a recently introduced applied science course which was assignment based. The assignments need to have vocational relevance. Students are required to meet specified performance criteria but teachers are given flexibility in the way this is achieved.

In discussions with a group of teachers and representatives from the awarding bodies, proficiency testing (PT) schemes were mentioned as part of the requirement for accredited laboratories. Teachers thought this could be used in the applied science course particularly as one of the mandatory modules is entitled 'Laboratory Safety and Analysis of Samples.' The pilot PT scheme for schools and colleges was therefore planned with these factors in mind.

Slide 2

So what is a PT scheme? A PT scheme is a means of assessing the quality of analytical measurement made in laboratories. The tests are 'round robin' exercises. The scheme comprises the regular distribution of homogeneous test materials to participating laboratories for independent tests. The results are returned to the organiser of the scheme who makes an analysis of the results and reports them to all of the participants.

The primary functions of the scheme are:
- to detect shortcomings in participants test procedures
- to provide feedback on any problems encountered.

This is a particularly important function in analytical measurement, which is fraught with practical difficulties and prone to unsuspected errors.

Schools and colleges were invited to take part in a 'one - round' PT scheme. The scheme was advertised as a competition and provided an opportunity for students to take part in a friendly analytical challenge.

Slide 3

There are number of factors to consider when planning a PT scheme aimed at schools and colleges.

It is advisable to work with a partner when organising a scheme. The PT schemes organised by the VAM Education and Training team, for pre-university students, have been run in conjunction with Nuffield Science (an awarding body responsible for a number of UK science qualifications at the 16 - 19 year old age range).

It is important to choose an analysis that can be easily performed by the centres participating in the scheme. The analysis chosen, in this instance, was an acid - base titration since most schools and colleges have access to the relevant equipment and the analytical technique (titration) is part of the teaching syllabus.

The sample or test material that is chosen to be analysed must be homogeneous, stable, cheap to produce or purchase, easy to distribute and pose no hazard in transportation or in use.

There are a number of key events in a PT scheme and these have to be scheduled to fit in with your partner and with the schools/colleges participating in the scheme. It is important not to clash with major events in the schools/colleges calendar, e.g. external examinations or school holidays. Also a selection of test samples will need to be tested before distribution to participating centres and so the timing of this laboratory work will need to be considered.

Slide 4
The points that have been raised will now be considered in greater detail.

Why work with a partner when organising a PT scheme? The obvious response is that the work load and costs associated with the scheme can be spread, however it was also a requirement that, if possible, events organised under the VAM programme are organised with a partner. Although a partner helps to lessen the work load it is still important not to be too ambitious on the number of centres invited to participate in a PT scheme. The scheme for schools/colleges with only 32 centres participating required many hours of planning for the exercise to run smoothly. The number of centres invited to participate will depend on the financial budget allocated to the scheme (purchase/preparation of test samples, postage and packaging, staff wages...) and the number of 'man-days' that can be spared for the organisation of the scheme. Therefore, money and time will be the factors that limit the size of the scheme. After planning a PT scheme, a budget should be produced in order to ascertain if the exercise is feasible.

A partner can also offer advise on the viability of the PT scheme as well as help decide on whether the exercise/analysis used in the scheme is appropriate. Nuffield Science (VAM partner in scheme) is a well respected organisation in the UK education community with considerable experience in joint activities. They were able to determine the relevance of the exercise and their involvement added credibility to the scheme.

The many contacts that Nuffield Science had previously developed (with schools/colleges) made it easier to select and invite centres to take part in the scheme.

There are times in a PT scheme when it is important that the identity of participating centres remains confidential, e.g. when analysing submitted scripts it is important for the titles/names of the centres to remain unknown to the assessor. This allows the assessor to remain impartial in his/her judgements and so there could be no claims of bias. Nuffield Science received all the submitted scripts from the centres. The scripts were then coded and sent to the VAM Education and Training team for analysis - in this way a centre's true identity was unknown when the submitted results were being assessed.

Slide 5
In the UK, two routes can be taken by students remaining in education after the age of sixteen.
- A vocational route - such as a General National Vocational Qualification (GNVQ). This is the applied science course referred to earlier. These courses are practically based and they are linked with industry.
- An academic route - such as Advanced Level (A Level) qualifications. These courses are more theoretically based but still contain practical components.

Therefore the PT scheme had to be designed to complement both these courses.

Slide 6
The PT scheme acted as support material for the relatively new GNVQ science

course and also for the more established A Level science courses. The PT scheme was suitable to both forms of qualification since it allows schools/colleges to compare their results:
- with an external standard of quality
- with the performance of their peers.

Slide 7
The analysis chosen for the PT scheme had to be relevant for both GNVQ and A Level science since these courses were run by the centres participating in the scheme. However, in the first instance, a PT scheme should be designed around a chemical analysis where a quantitative result is produced that can be expressed as a concentration (e.g. mol dm^{-3}) or as some other chemical measurement on a continuos scale (e.g. pH). Participants in the PT scheme, organised for schools and colleges, had to determine the concentration of ethanoic acid samples by titrimetric analysis. This practical exercise had the following advantages:
- The analytical technique used is relevant to both GNVQ and A Level science courses - mentioned in syllabus statements.
- Centres had appropriate apparatus and reagents to perform the analysis, therefore only the test material had to be distributed since no additional specialised equipment was required by the centres.
- The exercise could be completed within 1 or 2 lessons (i.e. 1 - 2 hours). To make the scheme a viable proposition to the centres it had to fit into a school's daily time-table. The demands on schools resources and teachers time by modern science courses make extra curriculum activity, such as PT schemes, unattractive unless they are self contained and take a relatively short time to complete.
- The test material (ethanoic acid) was homogeneous and stable over the time period when it would be analysed in the centres.

Note: Professional PT schemes do not assign a set method for determining the amount of test material, the choice of method is left to the discretion of the participant. Also, the concentration of the acid samples were accurately determined before they were distributed to the centres

Slide 8
The choice of a test material for a PT scheme will not simply depend on its homogeneity and stability, however these are important points.

The test material should be homogeneous. The ethanoic acid was vialed and a number of randomly selected vials were tested. The selected samples were analysed (by titration - validated method) in order to determine their concentrations. The concentrations did not differ significantly (statistical analysis) and so the sample was deemed homogeneous.

Test materials must be sufficiently stable to remain effectively unchanged in composition when exposed to conditions of storage, distribution and the elapsed time from the receipt to analysis of the material. Consequently, selected test materials will need to be evaluated for their physical and chemical stability. The centres analysing the ethanoic acid samples were given a three month period in which to complete the exercise, therefore stability testing of the test material had to be organised during this time period. Three months may seem a long time to allow for the analysis to be performed but this provided greater flexibility for participating centres to incorporate the activity into their teaching time-tables.

As well as being stable and homogenous, the test material should be relatively cheap to prepare or purchase since a number of samples will be required. Samples will be needed:
- for homogeneity testing
- for stability testing
- for analysis by participating centres
- as spares for use in the event of any breakage in transportation to the centre (allow three spare samples for each centre).

Centres participating in the PT scheme for schools/colleges were each sent five samples of ethanoic acid, each of slightly different concentration. Some centres wanted to enter a large number of students and so, in this instance, one acid sample per centre would have been inadequate. Therefore, this further increased the number of samples needed. Students were encouraged to work in teams to carry out the analysis - each team taking the role of a separate laboratory.

The container the test material is stored in should not react with the test material. The acid samples were sent to the participating centres in sealed glass ampoules. It was important that the glass ampoules were easy to package (not too bulky) and that they did not pose any hazard in transportation.

Slide 9

The timing and responsibilities for events in a PT scheme must be co-ordinated between the two organisations running the scheme. The following responsibilities are in the order needed for the PT scheme, aimed at schools/colleges, to run effectively:
- Nuffield Science and the VAM Education and Training team decided on an appropriate analyte, analysis, grading system and a plan was agreed for the scheme.
- The VAM Education and Training team obtained acid samples - to save extra expense it is advantageous if a company can be persuaded to donate test material.
- The VAM Education and Training team organised homogeneity tests and the first stability test on a selected number of acid samples - in order to determine suitability of the sample.
- Nuffield Science invited centres to participate - details as to the nature of the scheme were supplied.
- The two organisations jointly prepared student information sheets (practical details), student report sheet (reporting results) and teacher notes - it was important that the material was suitable for the intended audience.
- Nuffield Science collated a list of schools/colleges wishing to participate in the scheme. Each centre was assigned a code since confidentiality was needed when assessing submitted results.
- Nuffield Science distributed the paperwork, necessary to complete the analysis, to each participating centre.
- The VAM Education and Training team distributed acid samples to participating centres (Nuffield Science supplied names and addresses of the centres).

- The release date of the documentation and acid samples was co-ordinated so that the centres received them at the same time.
- The VAM Education and Training team organised stability tests at regular intervals over the duration of the scheme.
- Nuffield Science received the completed reports from the centres.
- Nuffield Science sent the completed reports to the VAM Education and Training team where the performances of the centres were assessed. The centres were only known as a code - this allowed for unbiased assessment.

How were the performances of centres determined? What criteria was used to define acceptable performance (data accuracy)? It was decided that statistical methods would be used to analyse and compare the results. The results were converted to a *Z score* which represents the closeness of individual results to an agreed value, i.e. the mean concentration of the acid samples determined from the homogeneity tests. This is the standard form of assessment used in professional PT schemes. However, unlike professional PT schemes - students were also assessed on the quality of the scripts which they returned (more information on assessment later).

Slide 10

A definite number of samples had to be tested for homogeneity before the samples were distributed to participating centres. So how many samples are needed for homogeneity testing?

The number of samples can be calculated using the formula:

$$\text{number of samples} = 3 \cdot \sqrt[3]{n} \tag{1}$$

where n is the total number of samples.

These are randomly selected from the bulk of the samples. Statistical analysis of the range of results from the homogeneity testing will determine whether the test material is homogeneous.

The stability of the sample will have to be monitored at regular intervals throughout the course of the scheme. In the case of the PT scheme for schools/colleges one stability test was performed every month on a randomly selected acid sample.

Slide 11

The paper-work sent to the centres consisted of:
- Student Information Sheets - these sheets explained the reasons why industry uses PT schemes and they also provided a general outline of how to perform the analysis.
- Student Report Sheets - these sheets allowed the students to explain how they used good laboratory practice to achieve satisfactory results. The report sheets were tabulated so that the students could clearly show their experimental data, any calculations performed and the final value they obtained (concentration of the acid sample).
- Teachers Notes – the reagents and apparatus needed for the analysis were outlined in these notes and the maximum number of students permitted to participate was outlined.

- Teachers Feedback Questionnaire - this allowed the teachers to comment on the relevance of the scheme and offer suggestions as to how to improve future schemes (further information via VAM (vam@lgc.co.uk)).

Slide 12

On receipt of the results, the organisers of the PT scheme can use statistical methods to analyse and compare results from participating centres. The results could be entered on to a spreadsheet where each result can be converted to a Z score which represents the closeness of individual results to an agreed value. These Z scores are then reported back to the participants. The statistical analysis highlighted which centres that had performed well. Scripts submitted from these centres were then graded on evidence of good laboratory practice.

Slide 13

As previously mentioned, the *Z score* is used as a measure of performance but how is it calculated.

From the equation it can be seen that the assigned value is taken away from the individual result. The assigned value, used in this instance, was calculated from the *mean* of the ethanoic acid concentrations determined in earlier homogeneity tests.

The numerator is calculated by subtracting the assigned value from the individual result, this is then divided by the 'established standard deviation', s, in order to calculate the *Z score*. However, how is the 'established standard deviation' found? It is a combination of the standard deviation of the homogeneity testing data and the standard deviation of the results from all participating centres (schools) i.e.,

$$S = \sqrt{(S_{schools})^2 + (S_{homogeneity})^2} \qquad (2)$$

The students taking part in this previous PT scheme were assessed on the concentration that they calculated for the acid sample. This incorporated not only any experimental error introduced but also assessed their ability to calculate the concentration of the acid from their experimental data. Earlier PT schemes simply assessed students on the experimental data that they submitted. Therefore the difficulty of the exercise can be determined by defining assessment criteria.

Slide 14

The *Z score* has been calculated for a particular team - but what does it mean?

If the modular *Z score* is less than or equal to 2 then this indicates that the result is satisfactory - participants will have performed with *distinction*.

A modular *Z score* between 2 and 3 acts as a warning of potential problems - participants will have performed with *merit*.

A modular *Z score* of greater than or equal to 3 acts as an indicator that there is a problem and that remedial action must be undertaken - individuals will have been judged to have *participated* in the scheme.

Slide 15

This overhead slide shows an example of a *Z score chart* - it shows the *Z scores* achieved by the participants in the scheme. centre is represented by a code and this ensures that any evaluation made is impartial. A centre is able to compare its performance with that of the other participants in the scheme.

Slide 16

The PT scheme for schools/colleges was run as a competition and so it was necessary to select a winner. In a professional PT scheme, centres are assessed and compared simply on the *Z scores* they achieve. However, in this instance, the different centres were firstly ranked by the *Z scores* they achieved. The scripts (report sheets) submitted by the students from the centres with the best *Z scores* were then evaluated. The scripts were graded on evidence of good laboratory practice and an understanding as to the sources and magnitude of errors in the analysis. In this way a judgement was made as to which was the best centre.

Slide 17

Due to the generally high standard it was necessary to select two centres as 'runners up' as well as the winning centre. The winning centres were awarded framed certificates. The winning centre received a cheque for £100 and the two 'runners-up' each received a cheque for £50.

Every student/participant in the scheme received a certificate. The grade of certificate (distinction, merit or participated) depended on the *Z score* achieved (see slide 14).

The PT scheme arranged for schools and colleges introduced the need for quality analytical measurements to young people studying science. The students were given the opportunity to match their skills against external competition. At the same time they gained a valuable insight into the nature of measurement in science. Because of its importance in the 'world of work' even having participated in the test/competition could be featured in applications for jobs and university places. The certificates, awarded to the students, could be used as part of their portfolio or 'Records of Achievements'.

The winning centres were awarded a framed certificate and cash prizes for their achievements. The prize awarding ceremonies were conducted with the support of the Royal Society of Chemistry (RSC) and an industrial partner in the neighbourhood of the winning centres. This helped these schools establish a link with industry and their Local Section of the RSC.

The centres were provided feedback as to the common problems encountered by the students in the analysis. Reference was made to practical skills as well as problems associated with calculations. It was the aim of the feedback to highlight aspects of the analysis that are critical in securing improvement in the quality of future analytical results. Each centre was provided with a *Z score* chart so that they could compare their performance with other centres.

With this aim in mind a resource material is being produced to encourage students to make valid measurements. It is entitled, 'A Guide to Improving Analytical Quality in Chemistry' and highlights what is considered best practice in basic analytical techniques such as titration, IR- and UV-spectroscopy.

Finally, the success of the PT schemes run for schools and colleges has also led to the production of a protocol for running a scheme. The protocol provides details as to how to set up a PT scheme for schools, for example, an university could use the protocol to help organise a PT scheme for its local schools.

Slides   159

Slide 1: Introduction
- The VAM programme
- Component for pre-university students
  - PT scheme linking academia and the workplace

Slide 2: Aims of a Proficiency Testing (PT) Scheme
- Benchmarking analytical laboratories
  - Test a known sample
  - Evaluation of results
- Provides feedback leading to improved quality

Slide 3: PT Scheme for Schools
- Select an appropriate partner
- What analysis to use?
  - Acid-base titration
- Choice of test material
  - homogenous, stable, cheap, non hazardous
- Time-table

Slide 4: Selection of Partner
- Why a partner?
  - Spreads the workload / requirement of VAM
  - Helps assess viability and nature of exercise
  - Selection of centres
  - Independence, need for confidentiality at certain stages

Slide 5: UK: 16+ Education
- Vocational Route - General National Vocational Qualifications (GNVQ)
- Academic Route A Levels

Slide 6: Support Material for Schools
- PT scheme allows the comparison of
  - results
  - performance

Slides 1-6

Choice of Test

- Fits in with syllabus
  - Common element titrations
- Availability of equipment
  - Advantages of using a titrimetric analysis

Test Material

- The test material must be homogenous
- The test material must be stable
- Number of samples must be considered
  – unit cost and number of samples required
- Packaging and distribution of samples

Allocation of Responsibilities
(J) = Joint (V) = VAM (N) = Nuffield

- Agree plan (J)
- Choice of test material (J)
- Obtaining test material (V)
- Homogeneity and stability testing (V)
- Choice of centres (N)
- Preparation of information pack (J)
- Distribution of
  – documentation (N)
  – test material (V)
- Receipt of results (N)
- Assessment (V)

Test Material Evaluation

- Homogeneity testing

  number of samples for testing = $3\sqrt[3]{n}$
  n = total number of units

- Stability testing

Dispatch Information Pack and Test Materials

- Information pack
  - student information sheets
  - student report sheets
  - teachers notes
  - feedback questionnaire
- Dispatch of test materials and information pack

Submission of Results

- Set completion date
- Receipt of results
- Enter results in spreadsheet
- Statistical evaluation of results
  - **Z scores** calculated
- Evaluation of reports

Slides 7-12

Slides   161

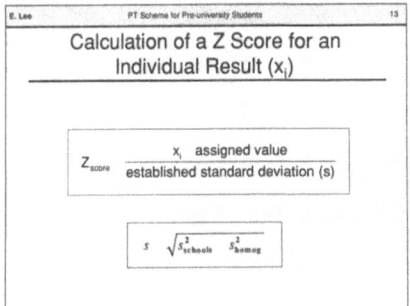

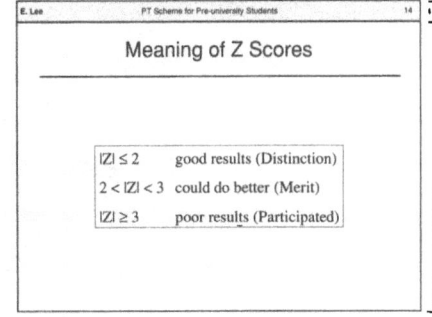

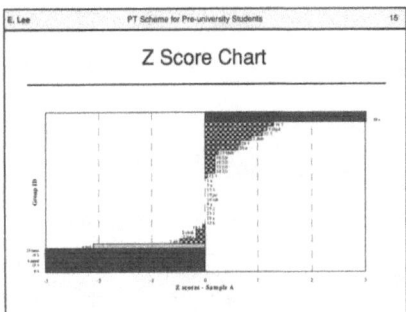

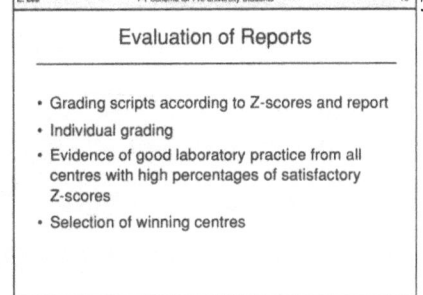

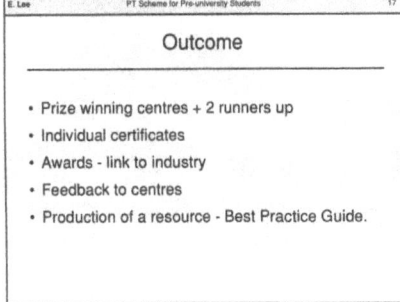

Slides 13-17

Teaching of the Concept of Valid Analytical Measurement: Integration of Quality Assurance (QA) Issues or Separate QA Courses for Higher Education

G. M. Greenway, A. Townshend

Abstract
The means by which the topic of Quality Assurance (QA) is introduced and taught in higher education courses, particularly in analytical chemistry, depends very much on the academic background of the 'students', the nature of the courses within which the teaching is to take place, and the timescale available. The choice between provision of dedicated QA courses or integration of QA principles into courses on other aspects of analytical science must be carefully considered. This presentation will address these aspects in order to help 'teachers' select the most suitable approach for their purpose. The discussion will be based mainly on experience at the University of Hull where QA concepts are taught to disparate groups of students, *viz.* conventional undergraduates studying analytical science, environmental science students, specialised analytical masters course students and those studying analytical chemistry by 'open learning'.

In all cases the students must be left in no doubt that analytical measurements should be valid and fit for purpose. They should be clear how such qualities are achieved and monitored, and should be convinced that such QA procedures are an integral part of analytical science. Of course, the discussion is equally relevant to other measurement disciplines, and some thought will be given to these wider connotations.

Slide 1
It is important that the principles of Valid Analytical Measurement (VAM) are understood by all analytical scientists. It follows, therefore, that analytical scientists must be knowledgeable in the procedures of Quality Assurance (QA) and Quality Control (QC), and have a commitment to implementing them. A major difficulty, however, is that analytical science often receives little attention in higher education courses, so that VAM issues are never discussed. Yet students are growing up in an environment where quality issues are becoming widespread, and quality is recognised as an important function for commercial success.

Slide 2
Teaching of quality issues in analytical chemistry, therefore, can involve three types of students:
- those that are being taught analytical chemistry as part of their higher education
- those that already have a considerable knowledge of analytical chemistry; because they have received substantial training, and may be currently employed as analytical chemists

- those that have not been taught any significant amount of analytical chemistry, but find themselves carrying out analytical work.

The approaches to these types of students are necessarily different, and are discussed below.

Conventional chemistry students

It is extremely desirable that all graduating chemistry students (B.Sc. level) should have a knowledge and understanding of modern analytical chemistry. About a third of those graduates who take up jobs as practising chemists become involved in analytical work, and many more will frequently make analytical measurements, so proper education in analytical chemistry would seem to be essential. Surprisingly, this is often not the case. Thus it is first necessary, if VAM principles are to be instilled into chemistry students that their undergraduate courses should have a significant analytical content. If they do not, students will require considerable further training after graduation, in analytical measurement generally, and, of course, in QA and QC procedures. The absence of analytical tuition necessarily and unfortunately reflects poorly on higher education institutions in the eyes of industrial employers.

Slide 3

Chemistry students attending analytical courses

There are, of course, higher education B.Sc. courses that include analytical chemistry to some extent. The so-called Eurocurriculum, and its exposition in the recently published textbook "Analytical Chemistry. The Approved Text" [1] has indicated the range of topics that might be included, and some Universities provide courses where an appreciable amount of the curriculum is covered. The University of Hull is one such institution. It should follow, therefore, that Hull has a commitment to teaching the VAM principles, and the QA and QC procedures that service them, and this is indeed so. These aspects are gradually built in as the courses proceed, and they are taught very much as integrated parts of the courses, rather than being the subject of a special course. We believe this is important because VAM is a natural objective of analysis, and it is less interesting, and more difficult, to discuss VAM in the absence of the appropriate level of analytical knowledge and experience.

Slide 4

There are some other factors that must be taken into account when teaching VAM, some of which are obvious, but require stating. Firstly, the information must be interesting, relevant and *correct*. The teacher must be completely committed to VAM; a half-hearted approach is of little value. Because of the varied backgrounds of students, mathematical concepts should be introduced gently. One should remember that QA and QC are not concepts restricted to analytical measurements. They will come over best within an environment where quality generally is seen to be important.

Slide 5

The Hull B.Sc. experience

All *first year* chemistry students spend their first three weeks in the laboratory carrying out titrations and gravimetric analysis. This is intended to ensure the proper use of basic chemical equipment. The experiments are all analyses of samples of industrial relevance, to emphasise that the measurements made have important (including financial) consequences for the industry concerned. The concept of VAM is therefore introduced at this stage, including accuracy, precision, errors, standards, reporting of results and good laboratory practice. Hopefully, this basic training carries forward to other aspects of chemistry in subsequent courses.

Slide 6
The University of Hull has a specialist B.Sc. course in chemistry with Analytical Chemistry and Toxicology. This lasts for three years at the University, with the option of the students spending a year in industry between years 2 and 3. Half the students take up this option. Other students may work in industry during their summer vacations. We are particularly concerned that students spending such periods in industry already have a good basic analytical knowledge, including the principles of QA and QC. Therefore, the second year analytical students, as well as learning about basic analytical techniques such as spectrophotometry and chromatography, receive specific tuition in quality issues. The topics covered include the importance of sampling, choice of methodology/instruments/reagents, statistics, accuracy, precision, certified reference materials, calibration, blanks, recoveries, uncertainty.

Slide 7
The course is assessed by an assignment in which the student is invited to find several methods for a particular analysis (e.g. cadmium in natural waters), standard methods if possible. They select the 'best' method, giving their reasons, and describe how they would ensure it was valid for their sample. In practical sessions they are expected to apply these principles.

Slide 8
Students returning from industry have no doubt that QA and QC is taken very seriously in industry. All final year analytical students, as well as receiving extensive teaching in advanced analytical chemistry and chemometrics, receive further tuition in QA and VAM. They critically and publically review published research papers and quickly learn that the published word often has serious deficiencies and/or omissions. They apply VAM principles to their advanced practical experiments, their group project (where students again evaluate different methodologies for solving an analytical problem), and their research project.

Slide 9
In all these courses the teaching and training materials produced by the Laboratory of the Government Chemist (UK) under the VAM programme have been most useful.

Environmental science students

Environmental science undergraduates will study less chemistry than chemistry students, but the chemistry that they will study will probably contain a greater *proportion* of analytical chemistry than that in standard chemistry courses. This is understandable, in that it is very clear to all participants that much of the study of the environment involves analysis. It is very important, therefore, for these students also, that they appreciate the concept of VAM. Thus the training in analytical measurement given to such students should be fit to ensure that they are competent analysts, able to achieve valid measurements, and are fully conversant with QA and QC concepts. There is evidence that such students are more *au-fait* with this situation than traditional chemistry students! At Hull, we are developing analytical courses for 'environmental' students, and these will include VAM.

Slide 10
Masters course students
In the UK, chemistry graduates who have studied little or no analytical chemistry (and these are in the majority), as well as graduates from other scientific disciplines, have the opportunity to build up knowledge of analytical chemistry by attending a 1-year M.Sc. course in Analytical Science (or similar title) at a University. Such courses cover a wide range of analytical topics, with practical experience, and undertake a research project for the final 4-6 months. Such courses should, and often do, offer tuition in VAM, both by incorporation of a specific course on the subject, and by utilisation of the principles in other parts of the course. This includes a group practical project (as above), including cross validation with standard methods. Each group consists of three or four students.

Slide 11
Industrial analytical chemists
Many 'students' are employed in industry as analytical chemists, whilst continuing to undergo training, sometimes in specialised techniques, but commonly over a wide range of analytical topics. Such students may attend universities part-time (e.g. one day a week for 5 years at B.Sc. level) or, with the development of information technology, by 'distance' or 'open' learning packages. These latter students generally communicate with their teachers at a distance, and only appear at the University for specific, and usually practically based, courses. At Hull we have developed a VAM programme for Analytical Chemistry by Open Learning (ACOL) students as a practical session based on the ACOL book "Quality in the Analytical Chemistry Laboratory" [2], and the VAM/RSC video on uncertainty [3]. Students work through a QA/QC session, based on a detailed mock Quality Manual. They audit the functioning of a mock-up analytical laboratory, based on an internal audit protocol, including one simple method of analysis. The students, already familiar with employment in an analytical laboratory, thus achieved a good insight into the need for, and execution of, a quality audit, and a mechanism by which it is carried out.

Slide 12
The mock audit is governed by a quality statement, against which the audit is executed.

Slide 13
The reporting structure of the hypothetical company is provided.

Slide 14
A timetable for audit is also given.

Slide 15
The training record for an analyst in respect of the analyses being audited is included.

Slide 16
A record of a particular sample is also given.
Further details of the mock audit can be obtained from Dr. Greenway [4].

Slide 17
Integrated versus separate courses
When taught as part of an analytical undergraduate course, VAM principles are best integrated with other aspects of the course. Time for study is limited, and competition for timetable slots by other subjects is always considerable. Thus an integrated approach is advantageous in that it is interesting, its relevance is apparent, and there is no consideration that it is a subject that is divorced from other aspects of analysis. However, the subject coverage is bound to be restricted, and the graduating student, although appreciative of the value of VAM, is by no means fully trained. A more comprehensive understanding of the principles and practice of QA and QC demands a specific course, aimed at 'students' who have a good background in other aspects of analytical chemistry. The content of such a course has been referred to above.
Non-Analytical VAM

The principles of VAM are by no means restricted to analytical chemistry. All aspects of measurement science will benefit from a proper appreciation of VAM, and the associated QA and QC procedures. Other branches of chemistry seem to be slow to receive this message. They cannot do other than benefit from it.

References
[1] R. Kellner, J-M. Mermet, M. Otto, H.M. Widmer, Eds., "Analytical Chemistry. The approved text to the FECS Curriculum Analytical Chemistry", Wiley-VCH, Weinheim, 1998.
[2] M.T. Crosby, J.A. Day, W.A. Hardcastle, D.G. Holcombe, R.C. Treble and E. Prichard, "Quality in the Analytical Chemistry Laboratory", Wiley, Chichester, 1995.
[3] "Confidence in Analysis" video and booklet, VAM/Royal Society of Chemistry, London.
[4] g.m.greenway@chem.hull.ac.uk

Teaching of the Concept of Valid Analytical Measurement:

Integration of Quality Assurance (QA) Issues or Separate QA Courses for Higher Education.

Important for all students

Industrial placement students

Analytical/environmental students

Reasons for Integrating the Principles of Valid Analytical Measurement (VAM) into Undergraduate Analytical Courses

- VAM is a natural objective of analysis
- VAM is best discussed in the context of an appropriate level of analytical knowledge and experience

---

Types of Students Involved

- Students being taught analytical chemistry as part of their higher education
- Those who have already received some training, and may be employed as analytical chemists
- Those who have received minimal training or education in analytical chemistry

---

Factors to Consider when Teaching VAM

- The information must be interesting, relevant and *correct*
- The teacher must be completely committed to VAM
- Mathematical concepts should be introduced gently
- QA and Quality Control (QC) are not concepts restricted to analytical measurements

---

Current Experience at Hull

First year (all chemistry students)

- 27 hour laboratory class
- titrimetry/gravimetry
- real applications
- emphasise good laboratory practice
- estimation of errors
- introduce accuracy, precision and standards

Introduction to Analytical Chemistry

Second year (analytical/environmental students)

- Importance of sampling
- Overview of methods
- Accuracy/precision/statistics
- Choice of instruments/reagents
- Certified reference materials
- Calibration
- Blanks/recoveries etc.
- Uncertainty

Slides 1-6

Slide 7: Second Year Assignment

- Each student assigned an analysis e.g. cadmium in natural waters
- Must find two or three methods (standard methods if possible)
- Select the one they consider "best", explaining why
- Go through method explaining how they would validate it

Slide 8: Students Returning from Industry

Quality Assurance (QA) Topics
Quality systems
Types of systems
Documentation
Quality control
Audit

Valid analytical measurements (VAM)
Sampling
Choice of method
Selection of equipment/materials
Measurement and calibration
Validation
Uncertainty (errors)

Slide 9: Materials Available

Laboratory of the Government Chemist

Tertiary Education Pack
Sampling examples
Case studies
Interlaboratory proficiency testing
Resource list

Videos
Lets agree to agree
Uncertainty
Certified Reference Materials

Computer packages
VAMSTAT

Slide 10: MChem/MSc Group Practical

- Each group assigned an analysis e.g. cadmium in natural waters
- Must find two or three methods (one standard method if possible)
- Each group member selects a method and validates that method
- A reference material is analysed by all methods and results compared
- The final report has a joint introduction and conclusions, but individual practical aspects

Slide 11: Industrial Analytical Chemists

Mock Audit (ACOL* students)
- Quality Assurance manual prepared
- Method set out in laboratory for analysis
- Students asked to audit laboratory with respect to method

Extension
- Method can be validated for a different sample

*ACOL: Analytical Chemistry by Open Learning

Slide 12: Mock Audit

Quality statement

This company aims to provide high quality analysis at reasonable costs for its customers. It achieves this aim by having planned and systematic actions and thus give customers the greatest confidence in our results.

Slides 7-12

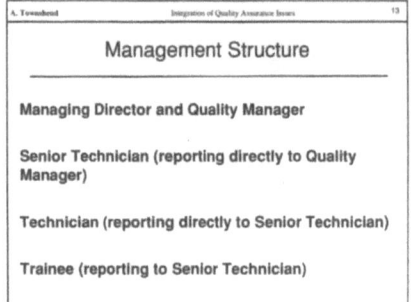
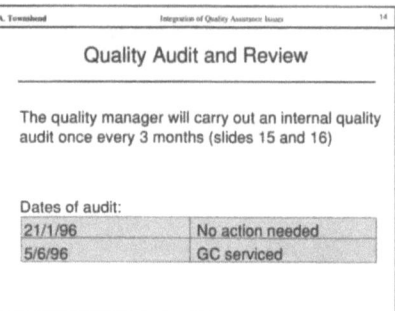
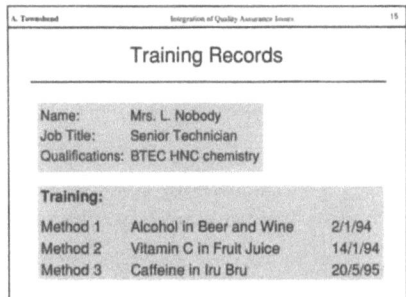
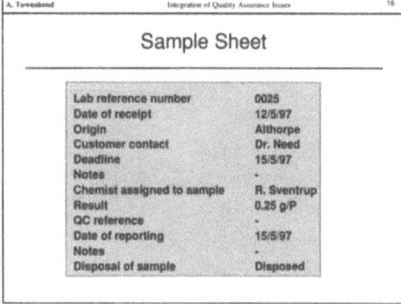

Slides 13-17

Special Requirements for Interlaboratory Proficiency Tests

M. Koch

Abstract
Interlaboratory tests are very important and very effective tools for proficiency testing. In order to fulfill this task they must meet some special requirements. Collusion between participants must be avoided, special efforts in the laboratories must be prevented, special attention must be paid to the definition of the assigned value and the quality requirements for the analysis must be stipulated.

In a proficiency testing scheme in Germany the University of Stuttgart developed a special design and evaluation method for interlaboratory tests that meet the above mentioned requirements.

Slide 1
The University of Stuttgart periodically organizes interlaboratory tests for the proficiency testing of analytical laboratories in the field of water analysis on behalf of the Ministry of Environment and Transportation of Baden-Württemberg with up to 250 participants. During this work, a special design and evaluation of these interlaboratory tests has been developed to fulfil the special requirements of interlaboratory tests for proficiency testing.

Slide 2
Interlaboratory tests are used for several purposes:
- evaluation and validation of newly developed analytical methods
- certification of reference materials in order to define the certified value with the highest possible accuracy
- proficiency testing for the supervision of analytical laboratories.

Slide 3
What are the special requirements for proficiency testing? What are the special problems?
Many laboratories have to successfully participate in proficiency tests in order to get contracts from their customers. So there is a link between the interlaboratory test and the commercial success of the laboratory. Thus there might be laboratories, which would like to provide a falsely optimistic impression of their capabilities. In order to avoid collusion between laboratories the organizer of a proficiency test should take preventive measures.

The organizer has to decide, how the assigned value should be defined. In contrast to interlaboratory testing for method validation or certification of reference materials, in proficiency testing in many cases less capable laboratories are involved. Therefore definition of the assigned value by consensus might be problematical.

Laboratories may also give a false impression of their performance if they routinely carry out single or double analyses, but report the mean results of many replicate analyses on the proficiency test samples.

The quality requirements that have to be fulfilled by the participants for successful participation must be defined.

Slide 4
As stated above, collusion between laboratories must be prevented. The only way to do this in an effective way is to distribute different samples. Nevertheless, the matrix and the difficulty of the analyses should be similar in all cases. The concentration levels should be different for each parameter, so that no one can conclude the concentration of one parameter, which possibly is more difficult to analyse, from the concentration of another one. Therefore, each laboratory should get an individual sample set.

Slide 5
There are several methods for the definition of the assigned value. The organizer of the interlaboratory test may select reference laboratories. The mean of the results of these laboratories could be defined as the assigned value, but even reference laboratories are not faultless. So there might be a bias.

One can use the mean of the results of all laboratories, but if there are many biased laboratories (e.g. analysis of highly volatile organic solvents in water) the assigned value will also be biased. For this reason synthetic or spiked samples should be used wherever possible, in order to obtain at least some independent information about the true value.

Slide 6
As stated above there might be a tendency in some laboratories to give a better impression of their capabilities by reporting the results of many replicate analyses, although in routine analyses they perform only a single or duplicate measurement.

To prevent this, the amount of sample should be limited. Of course it is a good practice in laboratories to perform duplicate measurement. Therefore at least two or three replicates should be possible.

Slide 7
Everybody knows that negative concentrations are not possible. A concentration cannot be less than nothing. Nevertheless, in most statistical calculations a normal distribution of the values is assumed. For low concentrations this implicates the finite possibility of negative concentrations. Therefore at low concentrations skewed distributions (e.g. log-normal distribution) must be used.

Slide 8
Interlaboratory test providers often distribute different concentration levels, either each level for each laboratory or a selection of levels for each laboratory.

If the evaluation is done level-by-level, the relative tolerance limits will vary with concentration. If there is a set of "good" laboratories the tolerance limits will be small compared with another level with a set of "inferior" laboratories.

Therefore, there will be a significant injustice resulting from the random assignment of the laboratories to the different concentration levels.

Slide 9
Scientific considerations will lead to the result that there should be a close connection between precision and concentration. The extent of the dependence could be different, but there should be no up and down.
However, how can this be achieved?
All concentration levels must be handled in one evaluation process.

Slide 10
For the proficiency tests in Baden-Württemberg a method has been developed for the normalization of all values (from all different concentration levels) to a single level. The formula used is a generalization of a formula developed by Horwitz.
In this formula the normalization coefficient describes the extent of the concentration dependence of the precision and therewith the tolerance limits.

Slide 11
If the normalization coefficient is equal to zero, there is no concentration dependence at all for the relative deviations from the assigned value.

Slide 12
With an increase of the normalization coefficient the concentration dependence also increases. A normalization coefficient equal to 1 would represent a constant absolute unprecision, but this is not likely to occur. This figure shows the concentration dependence of the precision and hence the tolerance limits for b=0.4.

Slide 13
This figure shows the empirically found normalization coefficients from many interlaboratory tests in Germany.

The parameters to be analysed are arranged in groups (inorganic non-metallic, metals, aromatic solvents, aggregate parameters, volatile halogenic solvents, pesticides and PAHs). For each group the range of normalization coefficients are depicted.

The boxes represent the range between the 10%- and 90%-quantiles, the whiskers the minimum and maximum values. Most of the parameters show normalization coefficients in the range between 0 and 0.25. Especially the organic parameters (analysed with chromatographic methods) have significantly lower values. This means that the concentration dependence of the uncertainty of these analyses is low. In contrast to this aggregate parameters as DOC, TOC etc. usually show a relatively high concentration dependence.

Slide 14
The purpose of proficiency tests is to separate laboratories with acceptable quality from laboratories producing unacceptable results. Therefore the organizer must define quality requirements for the laboratories. Maybe the quality requirements are defined by the customer. Then no statistical calculation is necessary at all. Nevertheless, the concentration dependence of the precision should be taken into account during the definition of quality requirements.

If there are no given requirements, these normally are calculated by statistical procedures. In contrast to interlaboratory tests for the certification of reference materials or for method validation, there are not only "good" laboratories involved in proficiency tests. The quality requirements should, however, be related only to the capabilities of "good" laboratories. Application of this procedure would require a pre-selection of "good" laboratories and the calculation of quality requirements (i.e. tolerance limits) only with the results of these laboratories.

Slide 15
In this figure the results of a laboratory from a proficiency test with 4 samples are depicted vs. the assigned values. Using these correlation points a straight line calculated by linear regression is drawn.

Slide 16
The straight line is characterized by two parameters: slope and correlation coefficient.
The slope is a measure for correctness of the calibration and hence the accuracy.
The correlation coefficient shows how precise the laboratory processed the samples. Therefore it's a measure for "intersample" precision.
A "good" laboratory (for this parameter) will have a slope and a correlation coefficient almost equal to one.

Slide 17
In this figure the "quality data" of 197 laboratories from a proficiency test for arsenic in water are shown. It can be seen that there is an accumulation of laboratories around the point 1.0/1.0. In this proficiency test these laboratories showed that they have high capabilities for the analysis of arsenic in water. In this sense these laboratories are thus "good" laboratories.

Slide 18
With this selection of "good" laboratories the tolerance limits can be calculated only from the results of laboratories with high capabilities, outlier tests are unnecessary. Laboratories with low quality have no influence on the tolerance limits.
For the definition of the assigned value this procedure can be used in an iterative way. A preliminary assigned value can be defined by reference laboratories, then the quality data of all laboratories are calculated and from the results of the good laboratories a new assigned value is calculated. This procedure can be repeated until there are no more changes.

Slide 19
There are some special requirements for proficiency tests as shown above. In this slide they are summarized.

Interlaboratory Tests

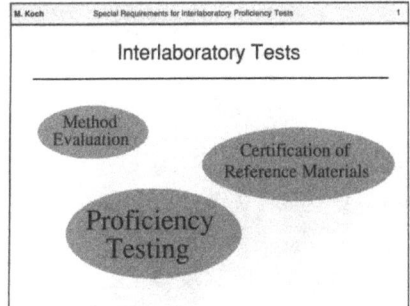

Problems for Proficiency Tests

- collusion
- fixing of the „assigned value"
- special effort in the laboratory
- definition of quality requirements

No Collusion!

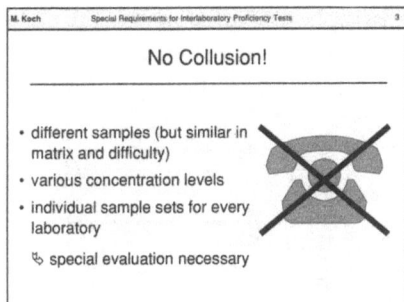

- different samples (but similar in matrix and difficulty)
- various concentration levels
- individual sample sets for every laboratory

    ↳ special evaluation necessary

Definition of Assigned Values

- by reference laboratories
  even reference laboratories may fail
- mean of all data
  may be shifted by biased laboratories

↳ spiked or synthetic samples (wherever possible)

No Special Efforts in the Laboratory

work in the laboratory should be done at routine level

    ↳ sample amount should be limited (but normal multiple analyses should be possible)

Statistical Distribution

- negative concentrations not possible
- at low concentrations skewed distribution

↳ e.g. log-normal distribution at low concentrations instead of normal distribution

Slides 1-6

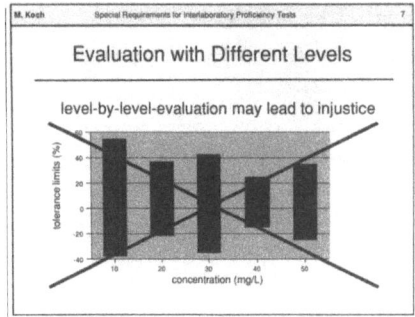
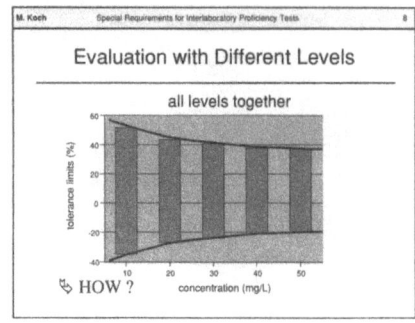
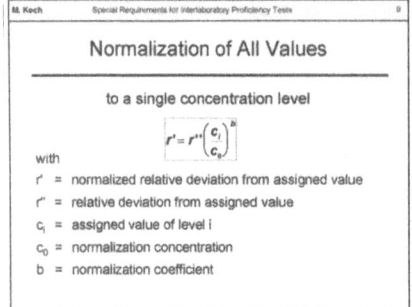
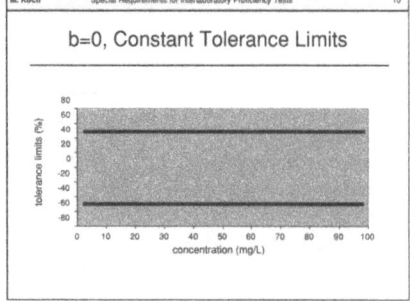
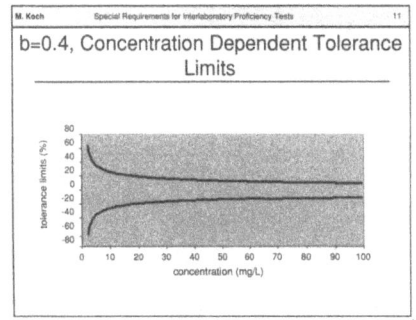
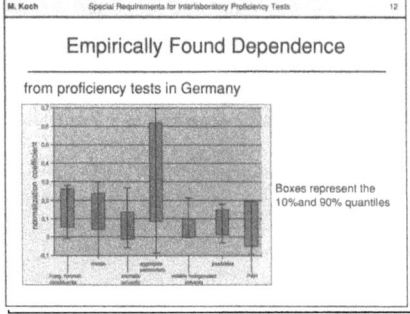

Slides 7-12

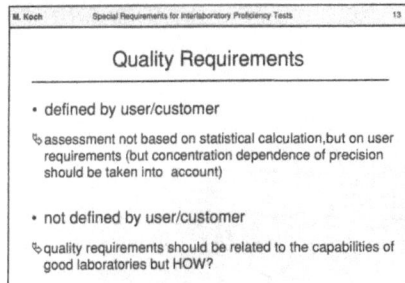
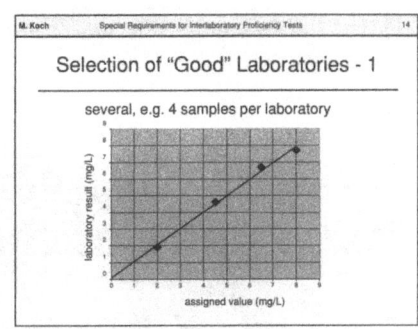
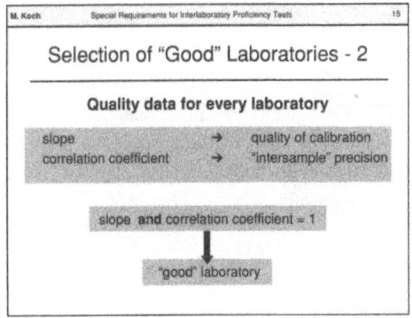
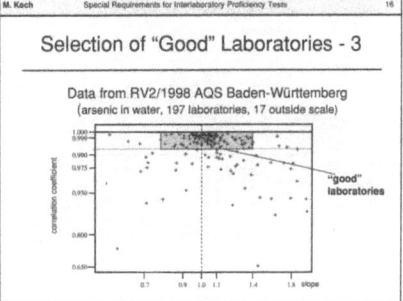

Slides 13-18

Location: http://www.springer.de/chem/

You are one **click** *away from a* **world of chemistry** *information!*

Come and visit Springer's
Chemistry Online Library

Books
- Search the Springer website catalogue
- Subscribe to our free alerting service for new books
- Look through the book series profiles

You want to order? Email to: orders@springer.de

Journals
- Get abstracts, ToC´s free of charge to everyone
- Use our powerful search engine LINK Search
- Subscribe to our free alerting service LINK *Alert*
- Read full-text articles (available only to subscribers of the paper version of a journal)

You want to subscribe? Email to: subscriptions@springer.de

Electronic Media
- Get more information on our software and CD-ROMs

You have a question on
an electronic product? Email to: helpdesk-em@springer.de

Bookmark now:

http://www.springer.de/chem/

Springer · Customer Service
Haberstr. 7 · D-69126 Heidelberg, Germany
Tel: +49 6221 345-217/218 · Fax: +49 6221 345-229
d&p · 006756_001x_1c

MIX
Papier aus verantwortungsvollen Quellen
Paper from responsible sources
FSC® C105338

If you have any concerns about our products,
you can contact us on
ProductSafety@springernature.com

In case Publisher is established outside the EU,
the EU authorized representative is:
**Springer Nature Customer Service Center GmbH
Europaplatz 3, 69115 Heidelberg, Germany**

Printed by Libri Plureos GmbH
in Hamburg, Germany